PRAISE for *Thinking 101*

"*Thinking 101* provides evidence-based advice that has real potential to improve lives."
—*Science*

"Ahn excels at illustrating how psychological concepts manifest in everyday life, and her suggestions provide sensible techniques readers can use to push back against cognitive biases. This heady volume provides plenty of food for thought."
—*Publishers Weekly*

"This book is not just a lucid overview of the cognitive traps that wreak havoc on your reasoning—it's also an expert's guide to rethinking how we think."
—ADAM GRANT, #1 *New York Times* bestselling author of
Think Again

"*Thinking 101* combines the best science with practical advice, to help you make better decisions. Ahn's stories are spot-on, they are humorous, and they show us how thinking can be turned on itself to overcome the biases from, well, thinking!"
—MAHZARIN BANAJI, professor of psychology at Harvard
University and coauthor of *Blindspot*

"Every day of our lives, we make judgments—and we don't always do a very good job of it. *Thinking 101* is an invaluable resource to anyone who wants to think better. In remarkably clear language, and with engaging and often funny examples, Woo-kyoung Ahn uses cutting-edge research to explain the mistakes we often make—and how to avoid them."
—GRETCHEN RUBIN, #1 *New York Times* bestselling author of
The Happiness Project and *The Four Tendencies*

"*Thinking 101* delivers a world-class tune-up for your brain. It will unclog your mental gears, restart your cognitive engine, and put you on the road to making smarter decisions."
—DANIEL H. PINK, #1 *New York Times* bestselling author of *The
Power of Regret*, *Drive*, and *A Whole New Mind*

T0050044

"There are other books on typical errors and biases of thinking, but Ahn's is remarkable. Not only does she limit her coverage to just eight major thinking problems, which allows her to deeply inform the reader about each with engaging, conversational prose, but she also offers compelling, research-based ways to limit the problems' unwanted impact. The result is a terrific one-two punch."
—ROBERT CIALDINI, *New York Times* bestselling author of *Influence* and *Pre-Suasion*

"*Thinking 101* is a must-read—a smart and compellingly readable guide to cutting-edge research into how people think. Building from her popular Yale course, Professor Woo-kyoung Ahn shows how a better understanding of how our minds work can help us become smarter and wiser—and even kinder."
—PAUL BLOOM, professor of psychology at the University of Toronto, Brooks and Suzanne Ragen Professor Emeritus of Psychology at Yale University, and author of *The Sweet Spot*

"Ahn's book is an absorbing, timely—and, I think, essential—guide to how our minds go wrong and what we can do to think better. With lots of humorous stories and cautionary thinking tales, this terrifically written book is a must-read for anyone who wants to understand and overcome the powerful yet invisible thinking traps that lead us astray."
—LAURIE SANTOS, professor of psychology at Yale University and host of *The Happiness Lab*

"Woo-kyoung Ahn uses wonderfully engaging examples to show how we can understand and improve our reasoning."
—ANNA ROSLING RÖNNLUND, *New York Times* bestselling coauthor of *Factfulness*

"*Thinking 101* breaks down when human thinking breaks down, and unlike many other books on the topic, this one is accessible, engaging, and fun to read. Woo-kyoung Ahn's delightful sense of humor shines through as she uses entertaining stories and examples to compellingly illustrate why thinking errors happen, why it matters, and what to do about it. The book is full of research-backed insights into how the mind works, which newcomers to the field will find clear and understandable, but also has a number of gems that more advanced readers will appreciate."
—DANNY OPPENHEIMER, professor at Carnegie Mellon University and coauthor of *Democracy Despite Itself*

Thinking 101

*How to Reason Better
to Live Better*

WOO-KYOUNG AHN

FLATIRON
BOOKS
NEW YORK

www.flatironbooks.com

Grateful acknowledgment is made for permission to reproduce art on page 206 from the following: Shali Wu and Boaz Keysar, "The Effect of Culture on Perspective Taking," *Psychological Science* 18, no. 7 (2007).

Charts courtesy of the author

The Library of Congress has cataloged the hardcover edition as follows:

Names: Ahn, Woo-Kyoung, author.
Title: Thinking 101 : why we so often get things wrong in life and
    how we can all do better / Woo-Kyoung Ahn.
Other titles: Thinking one-o-one
Description: First Edition. | New York, NY : Flatiron Books, [2022] |
    Includes bibliographical references and index.
Identifiers: LCCN 2022002313 | ISBN 9781250805959 (hardcover) |
    ISBN 9781250805966 (ebook)
Subjects: LCSH: Judgment. | Cognition. | Thought and thinking.
Classification: LCC BF447 .A46 2022 | DDC 153.4/6—
    dc23/eng/20220512
LC record available at https://lccn.loc.gov/2022002313

ISBN 978-1-250-80597-3 (trade paperback)

Our books may be purchased in bulk for promotional, educational, or business use. Please contact your local bookseller or the Macmillan Corporate and Premium Sales Department at 1-800-221-7945, extension 5442, or by email at MacmillanSpecialMarkets@macmillan.com.

First Flatiron Books Paperback Edition: 2023

10 9 8 7 6 5 4 3 2

To Marvin, Allison, and Nathan

CONTENTS

Thinking 101

INTRODUCTION

WHEN I WAS A GRADUATE STUDENT at the University of Illinois at Urbana-Champaign, doing research in cognitive psychology, our lab group went out every now and then for nachos and beers. It was a great opportunity for us to ask our advisor about things that wouldn't likely come up in our more formal individual meetings. At one of those gatherings, I summoned up the courage to ask him a question that had been on my mind for some time: "Do you think cognitive psychology can make the world a better place?"

I felt a bit like my question was coming out of left field; having already committed my life to this area of study, it was a little late to be asking it. But even though I had presented my findings at cognitive science conferences around the world and was on track to publish them in respected psychology journals, I had been having a hard time explaining the real-life implications of my work to my friends from high school. On that particular day, after struggling to read a paper in which the authors' primary goal appeared to be to show off how smart they were by tackling a convoluted problem that didn't exist in the real world, I finally found the courage to raise that question—with some help from the beer.

Our advisor was famous for being obscure. If I asked

him, "Shall I do A or B for the next experiment?," he would either answer with a cryptic "Yes," or turn the question around and ask, "What do you think?" This time I had asked him a simple yes-or-no question, so he chose a simple answer: "Yes." My lab mates and I sat there silently for what felt like five minutes, waiting for him to elaborate, but that was all he said.

Over the course of the next thirty or so years, I've tried to answer that question myself by working on problems that I hope have real-world applications. In my research at Yale University, where I've been a professor of psychology since 2003, I've examined some of the biases that can lead us astray—and developed strategies to correct them in ways that are directly applicable to situations people encounter in their daily lives.

In addition to the specific biases I've chosen to research, I've also explored an array of other real-world "thinking problems" that can cause issues for myself and those around me—students, friends, family. I saw how my students procrastinate because they underestimate the pain of doing an assignment in the future as opposed to doing exactly the same thing right now. I heard from a student who told me about a doctor who misdiagnosed her because he only asked questions that confirmed his original hypothesis. I noted the unhappiness of people who blame themselves for all their troubles because they only see one side of reality, and the unhappiness caused by other people who never see themselves as being at fault for anything at all. I witnessed the frustration of couples who thought they were communicating with perfect clarity but actually were completely misunderstanding each other.

And I also saw how "thinking problems" can cause troubles that go far beyond individuals' lives. These fundamental errors and biases contribute to a wide range of societal issues, including political polarization, complicity in climate change, ethnic profiling, police shootings, and nearly every other problem that stems from stereotyping and prejudice.

I introduced a course called "Thinking" to show students how psychology can help them recognize and tackle some of these real-world problems and make better decisions about their lives. It must have filled a real need, because in 2019 alone, more than 450 students enrolled in it. It seemed they craved the kind of guidance psychology could provide, and they told one another about it. I then noticed a curious thing: when I was introduced to students' family members who were visiting campus, they would often tell me how the students in my course would call home to talk about how they were learning to handle problems in their lives—and that some had even started advising other family members, their parents included. Colleagues told me they overheard students in the dining halls fiercely debating the implications of some of the experiments the course covered. When I would talk to people outside the profession about the issues discussed in the course, they asked me where they could learn more. All of this suggested that people really wanted and needed these kinds of tools, so I decided to write a book to make some of these lessons more broadly available.

I selected eight topics that I found most relevant to the real-life problems that my students and others (including myself!) face day to day. Each chapter covers one of them, and while I refer to material from throughout the book when relevant, the chapters are written so they can be read in any order.

Although I talk about errors and biases in thinking, this book is *not* about what is wrong with people. "Thinking problems" happen because we are wired in very particular ways, and there are often good reasons for that. Reasoning errors are mostly by-products of our highly evolved cognition, which has allowed us to get this far as a species and to survive and thrive in the world. As a result, the solutions to these problems are not always easily available. Indeed, any kind of de-biasing is notoriously challenging.

Furthermore, if we are to avoid these errors and biases, merely learning what they are and making a mental note that we should not commit them is not enough. It's just like insomnia; when it happens, you clearly know what the problem is—you can't sleep well. But telling insomniacs that they should sleep more will never be a solution for insomnia. Similarly, while some of the biases covered in this book may already be familiar to you, we need to provide prescriptions that are better than simply saying, "Don't do that." Fortunately, as a growing number of studies attest, there are actionable strategies we can adopt to reason better. These strategies can also help us figure out which things we can't control, and even show us how solutions that might initially seem promising can ultimately backfire.

This book is based on scientific research, mostly from other cognitive psychologists but also on some that I carried out myself. Many of the studies I cite are considered classics that have stood the test of time; others represent the latest results from the field. As I do in my course, I give a variety of examples taken from widely different aspects of our lives to illustrate each point. There's a reason for that, and you'll find out why.

So, back to the question I asked my advisor: "Can cognitive psychology make the world a better place?" In the years since I first posed it, I've come to believe ever more strongly that the answer is indeed, as my advisor so aptly replied, "Yes." Absolutely yes.

1

THE ALLURE OF FLUENCY

Why Things Look So Easy

WITH 450 SEATS, LEVINSON AUDITORIUM is one of Yale University's largest lecture halls, and on Mondays and Wednesdays between 11:35 and 12:50, when my undergraduate course titled "Thinking" meets, nearly every seat is filled. Today's lecture on overconfidence is likely to be especially entertaining, as my plan is to invite some students to come up to the front and dance to a K-pop video.

I begin my lecture with a description of the above-average effect. When one million high school students were asked to rate their leadership abilities, 70 percent assessed their skills as above average, and 60 percent put themselves in the top tenth percentile in terms of their ability to get along with others. When college professors were polled about their teaching skills, two thirds rated themselves in the top 25 percent. After presenting these and other examples of overly generous self-assessments, I ask the students a question: "What percentage of Americans do you think claimed they are better than average drivers?" Students shout out numbers higher than any of the ones they've seen so far, like 80 or 85 percent, giggling because they think they are so outrageous. But as it turns out, their guesses are still too low: the right answer is in fact 93 percent.

To really teach students about the biases in our thinking, it's never enough to simply describe results from studies; I try to make them experience these biases for themselves, lest they fall prey to the "not me" bias—the belief that while others may have certain cognitive biases, we ourselves are immune. For example, one student might think that he is not overconfident, because he feels insecure sometimes. Another may think that since her guesses about how she did on an exam are generally close to the mark, she is similarly realistic when she assesses her standing with respect to her peers in leadership, interpersonal relationships, or driving skills. This is where the dancing comes in.

I show the class a six-second clip from BTS's "Boy with Luv," a music video that has garnered more than 1.4 billion views on YouTube. I purposely chose a segment in which the choreography is not too technical. (If you've already found the official music video, it's between 1:18 and 1:24.)

After playing the clip, I tell the students that there will be prizes and that those who can dance this segment successfully will win them. We watch the clip ten more times. We even watch a slowed-down version that was especially created to teach people how to dance to this song. Then I ask for volunteers. Ten brave students walk to the front of the auditorium in a quest for instant fame, and the rest of the students cheer loudly for them. Hundreds of them, I am sure, think that they can do the steps too. After watching the clip so many times, even I feel like I could do it—after all, it's only six seconds. How hard could it be?

The audience demands that the volunteers face them, rather than the screen. The song starts playing. The volunteers flail their arms randomly and jump up and kick, all at wildly different times. One makes up completely new steps. Some give up after three seconds. Everybody laughs hysterically.

THE FLUENCY EFFECT

Things that our mind can easily process elicit overconfidence. This fluency effect can sneak up on us in several ways.

Illusion of Skill Acquisition

The class demonstration involving BTS was modeled after a study on the illusion of fluency that can occur when we are learning new skills. In the study, participants watched a six-second video clip of Michael Jackson doing the moonwalk, in which he seems to be walking backwards without lifting his feet off the floor. The steps do not seem complicated, and he does them so effortlessly he doesn't even appear to be thinking about them.

Some participants watched the clip once, while others watched it twenty times. Then they were asked to rate how well they thought they could do the moonwalk themselves. Those who watched the video twenty times were significantly more confident that they could do it than those who watched it just once. Having seen it so many times, they believed they'd memorized every little movement and could easily replay them in their heads. But when the moment of truth arrived and the participants were asked to actually do the moonwalk, there was absolutely no difference between the two groups' performances. Watching Michael Jackson perform the moonwalk twenty times without practicing did not make you a better moonwalker than someone who had only seen him do it once.

People often fall for the illusion that they can perform a difficult feat after seeing someone else accomplish it effortlessly. How many times have we replayed Whitney Houston's "And A-I-A-I-O-A-I-A-I-A will always love you" in our heads, thinking that it can't be that hard to hit that high note? Or attempted to create a soufflé after watching someone make one on YouTube? Or started a new diet after seeing those before and after pictures?

When we see final products that look fluent, masterful, or just perfectly normal, like a lofty soufflé or a person in good shape, we make the mistake of believing the process that led to those results must have also been fluent, smooth, and easy. When you read a book that's easy to understand, you may feel like that book must have also been easy to write. If a person hasn't done any figure skating, she may wonder why a figure skater falls while attempting to perform a double axel when so many others pull it off so effortlessly. It's easy to forget how many times that book was revised, or how much practice went into those double axels. As Dolly Parton famously said, "It costs a lot of money to look this cheap."

TED Talks provide another great example of how we can be misled by fluency. These talks are typically eighteen minutes long, which means their scripts are only about six to eight pages. Given that the speakers must be experts in their topics, one might think that preparing for such a short talk would be a piece of cake; perhaps some speakers simply wing it. Yet according to TED's guidelines, speakers should dedicate weeks or months to prepare. Speaking coaches have provided more specific guidelines for TED-style talks—at least one hour of rehearsal for every minute that you speak. In other words, you need to rehearse your talk sixty times. And those twenty or so hours are just for rehearsals—they don't include the hours, days, and weeks that go into figuring out what to include in those six to eight pages of script, and even more importantly, what you should leave out.

Short presentations are actually harder to prepare for than long ones, because you don't have time to think about your next sentence or feel your way toward the perfect transition. I once asked a former student who was working at a prestigious consulting firm whether he thought Yale had prepared him for his job. He said the one thing he wished he'd learned was how to convince a client of something in three minutes. That's the hardest kind of presentation to pull off because every word counts—but it looks so easy when it's done right.

Illusion of Knowledge

The fluency illusion isn't limited to skills like dancing, singing, or giving talks. You see a second type in the realm of knowledge. We give more credence to new findings once we understand how those findings came about.

Consider duct tape, for example. We use it to fix nearly everything, from patching a hole in a sneaker to making an emergency hem in a pair of pants. Studies have found that duct tape can also remove warts as well as or sometimes even better than the standard therapy of liquid nitrogen. It's hard to believe, until you hear the explanation: warts are caused by a virus, which can be killed when it's deprived of air and sunlight. Cover a wart with duct tape, and that's exactly what happens. Given this explanation of the underlying process, the therapeutic power of duct tape sounds that much more credible.

Some of my earlier studies were about this sort of phenomenon: namely, that people are more willing to derive a cause from a correlation when they can picture the underlying mechanism. Even though the actual data remains the same, we are much more willing to leap to a causal conclusion when we can envision the fluent process by which an outcome is generated. There's no problem with that, unless the underlying mechanism is flawed. When we are wrongly convinced that we understand a fluent process, we are more likely to draw a flawed causal conclusion.

Let me give you a specific example. While pursuing this line of research, I came across a book entitled *The Cosmic Clocks: From Astrology to a Modern Science,* which was written in the 1960s by a self-styled "neo-astrologer" named Michel Gauquelin. The book began with a presentation of statistics (although some of them are questionable, for the sake of this illustration, let's just assume they are all true). For example, Gauquelin says that those who were born immediately after the rise and culmination of Mars—whatever that means— are more likely to grow up to be eminent physicians, scientists, or athletes. He had hundreds or sometimes thousands of data points and used sophisticated statistics to draw his conclusions. Nonetheless, there were skeptics. Even he was puzzled by his own discoveries and searched for an explanation. He dismissed the less scientific hypothesis that planets somehow bestow certain talents on babies at the moment of their birth. He instead offered a seemingly fluent explanation. To some extent, he wrote, our personalities, traits, and intelligence are innate, which means they are already present within us when we are in utero. Fetuses send chemical signals when they are ready to be born, precipitating labor. And fetuses with particular personality traits signal when they are ready for labor in response to subtle gravitational forces that are determined by extraterrestrial events. Given such an elaborate explanation, even a skeptic may err by switching their response from "no way" to "hmm."

Perhaps the illusion of knowledge explains why some conspiracy theories are so persistent. The theory that Lee Harvey Oswald assassinated John F. Kennedy because he was a CIA agent may seem far-fetched, but when an additional explanation is added—that the CIA was concerned about the way the president was handling communism—it sounds more plausible. QAnon's theory that President Trump was secretly fighting a cabal of satanic pedophiles and cannibals who were hidden in the "Deep State" was said to have come from a source, "Q," whose high-level security clearance gave him access to the inner workings of the government. Of course, none of it is true, but the illusion of knowledge Q created by sprinkling his posts with operational jargon convinced many of their veracity.

Illusion Arising from Something Irrelevant

A third type of fluency effect is the most insidious and irrational of them all. What I've described so far are the effects of perceived fluency on matters immediately at hand. The perceived fluency of an impending task makes us underestimate the difficulty of executing it. Descriptions of the underlying mechanisms behind certain claims make unacceptable assertions of fact seem more acceptable, even though those "facts" haven't changed. But our judgments can also be distorted by the perceived fluency of factors that are completely irrelevant to the judgments they cause us to make.

For example, one study examined whether the names of stocks can affect people's expectations for their performance in the market. Yes, there are fluency effects in names. At first, the researchers used made-up names that were created to be easy to pronounce (Flinks, Tanley) while others were less pronounceable (Ulymnius, Queown). Although participants received no other information, they judged that shares with the more pronounceable (that is, fluent) names would appreciate, while shares with the less pronounceable (that is, disfluent) names would depreciate.

They also looked at real share names (Southern Pacific Rail Corp. versus Guangshen Railway Co., for example) and followed the changes in share prices on the New York Stock Exchange. The easy-to-pronounce shares did better than the difficult-to-pronounce shares; if one invested in 10 most fluently named shares and 10 most disfluently named shares, the fluently named ones yielded a profit of $113, $119, $277, and $333 more than the disfluently named ones, after trading for one day, one week, six months, and one year, respectively.

Some readers might think this happened simply because the companies with disfluent names sounded more foreign to people trading in American stock markets. So, in the final study, the researchers looked at the pronounce-ability of companies' three-letter stock ticker codes. Some, such as KAR for KAR Global, are pronounceable as words, while others, such as HPQ for Hewlett-Packard, are not. Surprisingly, the companies whose ticker codes were pronounceable performed significantly better on both the New York Stock Exchange and the American Stock Exchange than companies with unpronounceable, disfluent codes. The relative fluency of their ticker codes should have absolutely nothing to do with their qualities as companies—it is completely arbitrary—but investors nonetheless valued companies with codes pronounceable as words more highly than those with unpronounceable ones.

In case you don't follow the stock market, let's talk about a surreptitious fluency effect created by internet searches. These days, you can google anything. But the downside of having access to expert information is that it engenders overconfidence; it makes people think they are more knowledgeable than they really are, even about topics they did not actually google.

Participants in a study were asked to answer questions like "Why are there leap years?" and "Why does the moon have phases?" Half the participants were told to search for the answers on the internet, while the other half weren't allowed to do so. Then, in the second part of the study, all of the participants were presented with a new set of questions, such as "What caused the Civil War?" and "Why does Swiss cheese have holes?" These questions were unrelated to the ones asked during the first part of the study, so participants who used the internet had absolutely no advantage over

those who hadn't. You would think that both sets of participants would be equally sure or unsure about how well they
could answer the new questions. But those who used the
internet in the first phase rated themselves as more knowledgeable than those who hadn't, even about questions they
hadn't googled. Having access to unrelated information was
enough to inflate their intellectual confidence.

ADAPTIVE NATURE OF
THE FLUENCY EFFECT

Even though I understand the fluency effect, I sometimes
fall prey to it. Once, I watched a forty-minute YouTube
video on how to groom a long-haired dog. After I spent
another fruitless forty minutes trying to groom my beautiful
Havanese, I disproved the American Kennel Club's claim
that "Havanese are just as cute no matter what hairdo you
give them."

I am also a sucker for gardening catalogs. Whenever I
see pictures of impeccably cared-for gardens, especially
vegetable gardens, I order enough seeds to cover an acre of
land, which I don't have, and sprout them using special indoor lights. For all the time and money I spend, I have little
to show for it. Last year I harvested a grand total of four
peppers and had kale salad three times. But it all looked so
easy in the catalogs!

I have been teaching and researching cognitive biases for

more than thirty years, yet I was still fooled by the YouTube groomer's effortless, fluent demonstration and those glossy photos of flourishing gardens. Isn't the whole point of learning about cognitive biases to be able to recognize and avoid them? If I'm really such an expert, then why am I not immune to them?

The answer is that we can be susceptible to cognitive biases even after we learn about them because most (or perhaps all) of them are by-products of highly adaptive mechanisms that have evolved over thousands of years to aid in our survival as a species. We can't just turn them off.

The fluency effect stems from a simple, straightforward rule used in what cognitive psychologists call "metacognition," which means knowing whether you know something, like if you know how to swim, or what a fixed-rate mortgage is. Metacognition is a very important component of cognition. If you don't know how to swim, you know not to jump into a deep pool, even if you need to cool off quickly on a hot day. If "fixed-rate mortgage" doesn't sound familiar to you, then you know you should find out what it is before you sign up for one. Metacognition guides our actions: knowing what we know tells us what to avoid, what to search for, or what to dive or not dive into. We can't live without it.

One of the most useful cues for metacognition is a feeling of familiarity, ease, or fluency. We are familiar with things that we know and that we can do. If I ask you whether you know Mr. John Robertson, you may say yes, no, or maybe, depending on how familiar that name sounds to you. When you find yourself in a rental car office in a foreign country that only has cars with manual transmissions in its lot, you can judge whether you still know how to drive one based on how familiar it feels to move your left foot on a clutch while shifting a gear lever with your right hand.

But familiarity is just a heuristic, a rule of thumb or a quick-and-dirty way of finding good-enough answers without applying much effort. For instance, to determine how much house a person can afford to buy, one well-known rule is the 28 percent mortgage rule; your monthly payment should not exceed 28 percent of your monthly pre-tax income. Heuristics don't guarantee perfect solutions. The 28 percent rule is just a rough guide, and ultimately whether you can afford to buy a particular house or one at all depends on host of other factors. Likewise, using familiarity or fluency to make metacognitive judgments is a shortcut we use for situations in which we cannot systematically validate what we know. We cannot possibly take a swimming test every time we need to judge whether we can swim, so we rely on our feelings of familiarity.

The problem is that a heuristic that works to our advantage most of the time can sometimes play havoc with us, as we just saw. A person can become highly familiar with moonwalking after watching a video twenty times, and that feeling of familiarity or fluency can mislead them into thinking that they know how to moonwalk themselves. Similarly, the process of planting seeds in soil, fertilizing and watering them, and then harvesting delicious, ripe vegetables is easy to imagine, which creates the illusion of having a green thumb, even for a professor who teaches cognitive biases.

Although the fluency or familiarity heuristic sometimes leads us astray, it is a very useful tool for reminding us what we actually do know. This may very well be why humans have come to rely on it—because the benefits of metacognition outweigh the costs of the illusions it sometimes causes. OK, that was dense and abstract, so to be more concrete, let's go over this again, using an analogy from a well-known visual illusion, as Daniel Kahneman, a Nobel laureate in economics, did in his famous book *Thinking, Fast and Slow*.

The images of the world that we see with our eyes are projected onto a flat screen called the retina, a layer of light-sensitive tissue at the back of our eyeballs. Because the retina is flat, the images our brains receive through it are two-dimensional. The dilemma here is that the world is 3D. To perceive the world in 3D, the visual system in our brains utilizes various cues. One is called linear perspective, which is when parallel lines appear to converge toward a single point in the distance, as shown in the figure. Our visual system automatically assumes that whenever we see two lines converging toward a vanishing point, an object closer to that vanishing point (line A in the figure) must be farther away from us than an object in the foreground (line B in the figure). Since we know that objects that are farther away from us appear to be smaller, when we see two identical horizontal lines placed in linear perspective, our visual system assumes that the one closer to the vanishing point must be larger. In fact, line A and line B are exactly the same lengths, but our visual systems "think" that A must be longer than it is. This is called the Ponzo illusion after Mario Ponzo, the Italian psychologist who first demonstrated it. You can verify with a ruler or your finger that A and B are exactly the same lengths, but even so, you will still see A as longer. Likewise, cognitive illusions, such as the fluency effect, can persist even after you understand they are illusions.

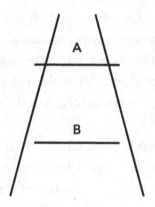

Furthermore, to say that we should always discount our feelings of fluency lest we become overconfident would be as absurd as saying that we should never use linear perspective but simply perceive our world as flat so we can overcome the Ponzo illusion. Illusions arise from the various cues and methods that our cognitive systems have adapted to allow us to navigate an uncertain world with infinite possibilities. Obviously, it's worth living with the Ponzo illusion if the system that causes it also allows us to perceive the 3D structure of the world. Likewise, it's better to be able to judge what we know or don't know by relying on our feelings of fluency, even if they occasionally lead us astray.

The analogy of visual illusions, however, stops there. Visual illusions rarely harm anyone. But overconfidence in the absence of sufficient evidence can have real-life consequences that are far more serious than temporarily ruining the appearance of a Havanese dog, or wasting fifty times more money on four peppers than what they would have cost at a grocery store. You could blow a career-changing presentation because you didn't adequately prepare for it, or lose your life savings by overrating the fluency of the name of a stock market share. You could invade the U.S. Capitol because you put too much faith in the stories you heard from QAnon.

But merely learning that fluency effects happen and are harmful isn't enough. It's like when a person gains unwanted weight: because our bodies are wired—for good reasons—to make us crave food, we need to go beyond merely thinking we should eat less and deploy concrete strategies to counteract that craving. So is it actually possible to sidestep the fluency effect despite our hard-wired metacognition? The answer is yes.

TRY IT OUT

While fluency effects stem from adaptations in our cognitive system, that doesn't mean we are powerless to overcome them. One simple solution is to make a task disfluent by physically trying it out. Read through your presentation out loud before you deliver it in front of an audience. Bake a soufflé first, before you invite your girlfriend's father for dinner. Sing "I Will Always Love You" to yourself in your bathroom mirror before you perform it at your company's holiday party in front of your boss. You won't need any feedback from others to break the illusion, because you will be giving it to yourself. I don't think any of those ten students who danced in front of the Levinson Auditorium still believe they could pull off a K-pop dance routine without practicing.

Trying out skills may seem like an obvious solution but surprisingly enough, not many of us actually do it. Some people think they're trying out skills when they're simply running the process in their heads and not using their physical muscles. When you imagine doing those dance steps or giving a presentation to your client, you are reinforcing the illusion. Everything flows smoothly in your mental simulation, feeding your overconfidence. You have to actually write down your presentation word for word and speak out loud using your tongue and vocal cords, or enact every movement of the dance using your arms, legs, and hips.

The importance of rehearsing is not limited to the acquisition of skills. We are frequently overconfident about the extent of our knowledge—we think we know more than we really do. A study demonstrated how spelling out one's knowledge can reduce overconfidence, even when no one gives us feedback. In this study, participants were first invited to rate how well they know how various things work, such as a toilet, a sewing machine, and a helicopter. On a seven-point scale, where 1 is "do not know at all" and 7 is "know exactly how it works," what rating would you give yourself for toilets, sewing machines, and helicopters? We are all familiar with these objects and have seen their parts operating smoothly. Even though we may not be able to build one from scratch, we kind of know how they work and what they do; in particular, we know how to flush a toilet. In the study, the average participant rated themselves in the middle, that is, around 4. That may not seem like much overconfidence, but it is and it's caused by the illusion of fluency.

To see for yourself, this time pick just one item, say, a helicopter, and write down or say out loud a step-by-step explanation of how it actually functions. Now rate yourself on how well you know how helicopters work. Most participants who were asked to do what you just did became significantly less certain about the extent of their knowledge. Trying to explain what they thought they knew was enough to make them realize how much less they knew than they'd assumed. One could go even further, as was done in the study, in which participants were probed with questions like "How does a helicopter switch from hovering to moving forward?" With each question, the participants became even more humbled.

This kind of a reality check can unfortunately occur *during* job interviews. You know the kinds of questions interviewers typically ask candidates: "Why are you applying for this job?" and "What are your strengths and weaknesses?" You think you know just what to say. Suppose the interviewer asks you, "What are your strengths?" You're glad they did, because you were prepared to say that your strengths are your organizational skills. Then the interviewer further probes you, asking, "Can you give us some examples?" Suddenly your brain freezes and you can only think of the last time you alphabetized your spice rack. Then they ask you questions like, "How would that help with *this* job?," and you realize that you won't ever have the chance to find out, because you're not going to be hired.

Talking through possible answers to practice questions is vital because you can objectify your responses. Once you write them out, you can pretend that they are someone else's and judge whether you'd give that person a job. You can also record yourself. I know, I know: it's excruciatingly painful to watch yourself on a video. Yet it's infinitely better that you see for yourself how well you can handle those questions before someone with decision powers does.

Along with the personal advantages of improving your presentations, job interview skills, and avoiding embarrassment at holiday parties, reducing your overconfidence can also help society at large. One study showed that it can reduce political extremism. Many of us hold strong views about various social issues, such as abortion, welfare, and climate change. Unfortunately, we may not realize how little we understand about them until we are asked to explain them.

In this study, participants were presented with various political policies, including imposing unilateral sanctions on Iran for its nuclear program, raising the retirement age for Social Security, establishing a cap-and-trade system for carbon emissions, and instituting a national flat tax. Participants were asked to state their positions on each of them, that is, how strongly they were for or against them. Then they were asked to rate how well they understood each policy's impacts.

Next, as in the earlier study on helicopters, participants were told to write down what those impacts are. After they finished, they were asked to re-rate their understanding of the policy. As before, their confidence was lowered. Having to explain what they knew in writing was all it took to make them realize how shallow their understanding was. So far, that is pretty much the same finding as in the helicopter study.

What was remarkable, though, was the last part of the experiment. At the end of the study, participants were asked to rate their positions on each policy again. It turned out that they became more moderate once their overconfidence was reduced. The more their illusion of knowledge was shattered, the less extreme they became. It's worth emphasizing that they did not moderate their views in response to opposing arguments. All it took was a little nudge to explain themselves.

That is why it is so important for society that we have

conversations with people who hold different views than we do. We tend to be drawn to people who share our views. When we stay in our bubbles, we do not talk about the impacts of the policies we support, because we assume that our allies already know them. It's only when we are forced to explain the consequences of the positions we hold to someone who does not share our views that we can begin to recognize the holes in our knowledge and the flaws in our reasoning, and work to fix them.

WHEN YOU CAN'T TRY THINGS OUT: THE PLANNING FALLACY

Unfortunately, there are many situations that don't allow us to dampen our overconfidence simply by trying out our skills or spelling out our knowledge explicitly. To explain this, we need to consider the planning fallacy.

We frequently underestimate the time and effort we need to complete a task, which is why we so often miss deadlines, exceed our budgets, or run out of energy before we finish. One of the most notorious examples of the planning fallacy happened when the Sydney Opera House was built. It was initially budgeted at $7 million, but ended up costing $102 million for a scaled-down version, while taking ten years longer to complete than originally estimated. The Denver International Airport overran its budget by $2 billion and took sixteen months longer to complete than estimated, which some people say is the reason there are so many conspiracy theories about it. One of those theories is that it took so long to build because it includes a network of secret underground bunkers in which billionaires and politicians can take refuge in the event of an apocalyptic event. Another conspiracy theory involves aliens, of course. The theories are so pervasive that the airport itself opened up a conspiracy museum. All that said, it would be unfair to New Englanders to overlook Boston's Big Dig highway construction project, which went $19 billion over its budget and took ten years longer to complete than projected.

The planning fallacy doesn't only apply to construction projects. The Standish Group, an independent international information technology (IT) research advisory firm, produces annual reports about various projects. You might think IT people would know how to use data from the past to make accurate predictions about the future. According to the Standish Group, however, between 2011 and 2015, the percentage of successful IT projects in the United States—with success defined as all required features completed on time, on budget—hovered between 29 and 31 percent. Half the projects were delivered late, over budget, or missing required features, and between 17 and 22 percent of projects simply failed. And there were no signs of a trend toward improvement.

The planning fallacy has several causes. One of them is wishful thinking: we hope our projects will be completed sooner rather than later, without our having to spend too much money on them, and these wishes are reflected in our planning and budgeting.

It's also important to recognize that the planning fallacy is in large part a type of overconfidence that stems from the illusion of fluency. When we are planning, we tend to focus only on how the project *should* run, on the things that have to be done to make it successful. When you picture those processes in your mind, they all run smoothly, engendering overconfidence.

One study that examined the planning fallacy revealed exactly this dynamic, while teaching us *what not* to do if we want to avoid it. Participants were asked to estimate how long it would take for them to finish their Christmas shopping. On average, they estimated they would be done by December 20. That turned out to be an illustration of the planning fallacy, because the participants didn't finish their shopping until December 22 or 23 on average.

To avoid succumbing to the planning fallacy, it might seem like a good idea to come up with specific, detailed plans. So another group of participants was instructed to write down step-by-step plans for their Christmas shopping. For example, a participant could make a list of family members and possible gifts for each of them. Another might select which mall to tackle on what day, and plan what to look for in that mall for each person on their list. While coming up with such plans, they looked totally executable. But did having them improve the quality of their estimates for how long it would all take? On the contrary, these participants displayed an even worse planning fallacy; most thought they would finish shopping seven and a half days before Christmas, which was three days earlier than the estimate made by those who didn't come up with step-by-step plans. But on average, they also finished around December 22 and 23.

The reason why having step-by-step plans exacerbated the effects of the planning fallacy was because the plans they came up with created the illusion that their shopping would be as smooth and effortless as in *Pretty Woman*, when Julia Roberts found all those perfect dresses in her size in a shopping spree that lasted less than half a day, or in *Clueless*, when Alicia Silverstone finished her shopping with her makeup showing no signs of wear while carrying two giant shopping bags down the street as if they weighed nothing.

Still, this is not to say that we should never come up with step-by-step plans. Breaking a task down into bite-sized steps, and meticulously setting deadlines for each of them, is an important part of planning, especially when the task at hand is more complicated than holiday shopping. A different study showed that when a task is divided into smaller subtasks, the planning fallacy is somewhat reduced. Unpacking a task forces people to realize that it isn't as simple as they imagined. Yet one should be aware that it can also create an illusion of fluency, inflating one's sense of control and aggravating the planning fallacy.

How can we counter that illusion? Earlier, I mentioned that to reduce overconfidence stemming from fluency, we can actually try things out. The irony with overcoming the planning fallacy, of course, is that the challenge is to plan things *without* trying them out. We can't practice Christmas shopping or building an opera house. But what we *can* do is make our mental simulation more disfluent by considering potential obstacles. There are two kinds of obstacles to consider, and one is more readily accessible to our minds than the other.

It's relatively easy to think of obstacles that are directly relevant to the task at hand. For holiday shopping, there will be traffic jams the weekend before Christmas, or stores might have sold out their stocks of that leopard-patterned cashmere cardigan you thought would be perfect for your grandmother. Such task-relevant obstacles would likely be factored into your planning.

What tends to be neglected, however, are the obstacles that have nothing to do with the task, such as coming down with a cold, your cat wandering off, your water heater leaking, your son breaking his ankle, et cetera, et cetera. Unexpected contingencies like those are difficult to plan for, simply because there are way too many possibilities. Furthermore, even if you remember that your son broke his ankle last year during the Christmas shopping week and you had to spend a whole day at the ER, you wouldn't expect that to happen again this year.

Unexpected events, however, are known unknowns. One thing about life that we know for certain is that there is always something. We just don't know what it is. My own solution, not based on any scientific evidence but from having personally experienced plenty of planning fallacies, is simple: I always add 50 percent more time to my initial estimate, like when I tell a collaborator that I can look over the manuscript in three days, even though I actually think I can do it in two. This strategy works fairly well for me.

OPTIMISM AND FLUENCY EFFECTS

While contemplating how we can avoid the fluency effect, it is also valuable to discuss the things that can magnify it, one of which is optimism. Optimism is like engine oil for the fluency effect: it makes everything appear to run more smoothly. When we are feeling optimistic, we shut our eyes to potential setbacks and obstacles.

Generally speaking, however, optimism is a good thing. Being optimistic can reduce our stress and make us feel happier. Feeling happy and having less stress improves our mental as well as our physical health, and probably because of that, optimists live longer. Not only is optimism good for our health, it is essential for our survival; we all know we will die eventually, so without some optimism about our futures, we wouldn't be able to motivate ourselves to pursue anything.

Some argue that optimism is particularly advantageous in competitive situations. Suppose Tom and Jerry are business rivals, bidding for the same construction projects all the time. Jerry's company is much smaller than Tom's, and Tom can almost certainly outbid Jerry. If Jerry is not optimistic and acts solely in accord with the objective truth, he will just give up. If Jerry is optimistic, however, he can at least carve out a place for himself by pursuing projects that Tom isn't interested in.

For these reasons, it's likely we are hardwired for optimism to a certain extent, as demonstrated by studies done with nonhuman animals like birds and rats. In one study, European starlings learned that when they heard a two-second-long sound, they should press a red lever to get food, but when they heard a ten-second sound, they should press a green lever to get food. If they pressed the lever that mismatched the sound, they wouldn't get any food at all. In addition, the researchers made the red lever a better deal than the green lever: pressing it delivered food immediately, while pressing the green lever delivered food after some wait. Nobody enjoys waiting for food. Having learned all these contingencies (and it is quite impressive that these little birds could do that!), the experimenters presented the birds with a devious test: this time, they played a medium-length, six-second tone. What would the birds do, press the red lever or the green lever? The birds were optimistic. Given the ambiguous tone, they pressed the red lever—the one that gave them the better option.

Because optimism is a default mode for most people, it can easily worsen the fluency effect, resulting in blind optimism. Realistic optimism is when you say that a glass is half full, or that there's a light at the end of a tunnel. Blind optimism is when you deny that the glass is half empty or that you are even in a tunnel. A historic example of blind optimism that is still vivid in everyone's minds is the way COVID-19 was handled in the United States during the critical first days and weeks after it emerged, when no federal-level measures were taken to prevent its spread. Some believed the virus would magically disappear in the spring, when there's more sunlight and the temperatures rise. Because a world with lockdowns, quarantines, no concerts, no vacations, and no restaurants for more than a year was unimaginable, and because it was easier to imagine a normal April after a flu season, many fell for blind optimism. Could that have been avoided?

One method that's been shown to be an effective deterrent to blind optimism is to make people think about similar cases in the past and earnestly apply their lessons to the current one. Simply thinking about similar experiences is vastly useful, but it's not enough. Even when those similar cases are brought to our attention, we tend to dismiss them, saying, "Oh, this time it's different," "I learned my lesson last time and it won't happen again," and so on. When we were first dealing with COVID-19, many people compared it to the 1918 pandemic (the Spanish flu), but the lessons from that pandemic were easy to disregard—"We have much more advanced medical knowledge today, and besides, that was a completely different virus." Even when we read about what was happening in China, it was tempting to think that America would be different, as if calling it "the Chinese virus" made us immune to it.

All this clearly illustrates why it's not enough only to think about similar cases if we are then simply going to focus on why this time it will all be different. To avoid the temptation to make excuses and fall for blind optimism, we should assume that the current case will be the same as past ones and make our plans and predictions accordingly. In the case of COVID-19, we should have assumed that the disease would spread in New York, L.A., or anywhere else in the world exactly as it did in Wuhan. Data-driven predictions are more accurate than predictions based on intuitions or wishful thinking.

SUMMING UP: REMODELING MY HOUSE

As a way to wrap up this chapter, let's talk about my plans to remodel my house and how what I've been discussing can be applied to make them go better. Our house is about a hundred years old, but without any old-house charms; we bought it because of its great location. Half of its windows either don't stay open or don't open at all. The second bathroom has 1960s features—namely, plastic and linoleum—that no fancy shower curtain or bath towels can hide. Pieces of its original siding fall off in wind storms, providing instant mulch for our garden. The most annoying thing to me, after having been confined in it every day for one and a half years because of the pandemic, is the half wall that divides the living room into two spaces for absolutely no reason. We've decided to knock it down.

Neither I nor my husband know much about how to maintain a house. When we bought our first house together twenty-five years ago, we asked the owner what to do with its casement windows when it rains; we were worried that their gorgeous wooden frames would be ruined. The owner, who had built the house, said, "You close them," and appeared quite nervous about selling the house to us. Given everything I've said in this chapter, however, my lack of confidence about remodeling is a huge advantage.

Although knocking down the half wall in the living room seems like a simple matter of sledgehammering, this could be yet another example of the fluency effect, spelling the end of our upstairs master bedroom as we know it. I've also picked a minimalist design for the second bathroom that doesn't appear too hard to re-create in our space—minimalism by definition means simple. But many experts on remodeling suggest that you allow for as much as 50 percent more time and money than the contractors' estimates, so we've done exactly that. As the windows get replaced, the contractors may find water damage, mold, wasp nests, and everything else that I refuse to believe exists in my house but need to be prepared for, both psychologically and financially. The last remodeling also taught me that I cannot leave the contractor alone too long, as he made some creative adjustments on his own. This time, I'm not going to leave the contractor alone for any great length of time. The remodeling won't be done as easily as turning the pages of *Architectural Digest*, but there will be a light at the end of the tunnel.

2

CONFIRMATION BIAS

How We Can Go Wrong
When Trying to Be Right

LATE ONE AFTERNOON, WHILE I WAS wrapping up some work in my office, I received a call from Bisma (not her real name), a former advisee and one of the brightest students to ever take my "Thinking" course. She sounded upset, and since she's not someone who gets frazzled easily, I dropped what I was doing and paid attention.

She told me that she'd just left the office of a new doctor. Bisma had suffered from mysterious health problems since she was in high school. She couldn't keep her food down, especially in the morning; sometimes her nausea was so bad she'd faint. As a result, she was very thin. Doctors had ruled out most of the usual suspects, such as celiac disease, ulcers, and stomach cancer, but they hadn't been able to figure out what was causing her symptoms. She'd gone to this new doctor, she told me, because she needed to renew a prescription for her anti-nausea medication before she headed off for a semester abroad in Nepal and Jordan. The doctor listened to her politely as she described her symptoms. Then he asked, "Do you enjoy throwing up?"

It was clear to Bisma that he suspected anorexia. She was so thrown off guard that she couldn't recall exactly how the rest of the conversation went, but as she reconstructed it, it went something like this:

Bisma: No, I don't enjoy throwing up.

Doctor (thinking, *Of course she denies her problem*): Do you enjoy food at all?

Bisma (wondering who on earth would enjoy food if they suffered from chronic digestion issues like hers): No.

Doctor (thinking, *That's exactly what I suspected. Now we are getting somewhere*) Do you want to kill yourself?

Bisma: No!

By that point, Bisma was so upset that she walked out of the consulting room. The doctor, who must have interpreted her reaction as hysterical denial, became all the more convinced of his diagnosis and assumed she was not just running out of his office but away from her issues. He followed her out into the waiting room and shouted at her in front of the other patients: "Come back to my office! You have a serious problem!" She rushed out to her car instead, and called me.

Bisma went off to her study abroad program, but it was canceled midsemester due to COVID-19. During the two months that she was away, her symptoms disappeared.

Nobody knows for certain what caused her nausea and weight loss, but Bisma now thinks it likely that she was allergic to something in the United States and that the time she'd spent away from the source of the allergens allowed her immune system to calm down. What we do know for certain is that she was never treated for anorexia, and that her stress levels certainly hadn't decreased with the onset of a global pandemic and the disruption of her plans for her junior year.

While we now know that her doctor's diagnosis of anorexia was wrong, we can also see why he would have been so confident about his diagnosis. Bisma was extremely thin; most other common causes for her symptoms had been ruled out; she'd told the doctor that she didn't enjoy food; and she exhibited an unusually strong denial of potential psychological problems. Where the doctor went off track was in only asking her about things that confirmed his suspicions, and asking them in such a way that they would stay confirmed, no matter how she answered.

WASON'S 2-4-6 TASK

Try this problem: I am going to give you a sequence of three numbers. The sequence is determined by a simple rule, which you need to figure out. Remember the rule is about the sequence—that is, the relationships between the three numbers. The way you'll figure it out is by giving your own sequences of three numbers to me. Each time you propose a sequence, I will tell you whether it follows the rule or not. You can test as many sets as you want. When you are sure you've figured out the rule, tell me. I'll then let you know if that's the rule I used to generate my sequence.

Ready? Here are the three numbers: 2, 4, 6.

What three numbers would you test? Let me illustrate what typically happens in an experiment using this problem. Say a student named Michael gives it a try and I'm the experimenter. Michael generates 4, 6, 8 and I tell him that his sequence follows the rule. Michael thinks he's got it. "This is too easy," he says. "The rule is even numbers increasing by two." I tell him that he's incorrect.

Michael revises his hypothesis. "OK, fine," he thinks, "perhaps it's not even numbers but any numbers increasing by two." Proud of figuring this out, he tests 3, 5, 7, thinking that the answer had better be yes. Indeed, I say "yes." Trying to be extra-careful, he now tests 13, 15, 17, and gets another yes. So, he triumphantly declares, "*Any* numbers increasing by two!!!" I tell him that's not the rule either. Michael got a perfect math score on his SAT, so this is really a blow to his ego. He tries again:

Michael: -9, -7, -5
Me: Yes.
Michael: Hmmmmmmm. OK, how about 1004, 1006, 1008?
Me: Yes.

Michael: Gosh, how could it not be any numbers increasing by two then?

Michael did what most participants in Peter Wason's famous 2–4–6 experiment do. He tested his hypothesis by collecting evidence that would confirm it. Confirming data is necessary but not sufficient, because you also need to *disconfirm* your hypothesis. To show you how to do that, let's start by revisiting the sequences that I said conformed to the rule. They were:

2, 4, 6
4, 6, 8
13, 15, 17
-9, -7, -5
1004, 1006, 1008

There are in fact an infinite number of possible rules that are consistent with these data. Numbers increasing by two that have the same amounts of digits. Numbers increasing by two that are greater than -10. Numbers increasing by two that are greater than -11. And so on, and so on.

We cannot test all these hypotheses, but the point is that when there are so many possible rules that can explain the available data, considering only the hypothesis that comes first to your mind won't allow you to find the correct one in this case.

Thinking along those lines, Michael decides to consider an alternative rule: "Numbers increasing by the same amount." To disconfirm his original hypothesis and test this alternative, he gives me 3, 6, 9. I say yes.

Michael: I got it. How about 4, 8, 12?

Me: Yes.

Michael (using a fancy equation to prove he's not dumb):

OK, I'm sure it's X + k where X is any number and k is a constant.

Me: No.

What Michael should do is to try to disconfirm his hypothesis again. Really frustrated, he generates a random sequence.

Michael: How about 4, 12, 13?

I smile and tell him yes—it conforms to the rule.

Michael: WHAAAT????

That was the critical test, one that violated the hypothesis he was testing at the moment. After pondering for a while, Michael says, "5, 4, 3?"

I shake my head no. That sequence does not follow the rule.

Considerably humbler at this point, Michael meekly offers, "Could the rule be any increasing numbers?"

I say, "YES! Exactly."

CONFIRMATION BIAS

Peter C. Wason was a cognitive psychologist at University College London. He devised the famous 2–4–6 task in 1960, providing the first experimental demonstration of what he called confirmation bias, our tendency to confirm what we already believe. Back in those days, almost all psychologists of reasoning assumed that humans are logical and rational. As one might expect from the psychologist who coined the term "confirmation bias," Wason disconfirmed this commonly held belief.

In Wason's first experiment with the task, only about one-fifth of participants found the correct rule without first announcing any incorrect rules. Wason was so shocked by how many people could not solve this apparently simple problem that he thought the issue might lie with the structure of the experiment itself and looked for ways to fix it. When the experiment was repeated at Harvard, participants were told they had only one shot at the correct answer. That, he hoped, would force them not to rush. Even so, 73 percent of the participants announced an incorrect rule.

Some even resisted, insisting, "I can't be wrong since my rule is correct for those numbers," or "Rules are relative. If you were the subject and I were the experimenter then I would be right." One participant didn't announce any rule but happened to develop psychotic symptoms during the experiment—who knows why—and had to be rushed to the hospital in an ambulance. Another participant came up with an impressive rule: "Either the first number equals the second minus two, and the third is random but greater than the second, or the third number equals the second plus two, and the first is random but less than the second." He elaborated on this for fifty minutes before giving up.

Keeping the 2–4–6 task in mind, now let's revisit Bisma's encounter with the doctor. He diagnosed her with anorexia and then asked only questions that would confirm that belief. As a result, all of his evidence was positive: a young woman who throws up frequently, is very thin, does not enjoy food, and overreacts to questions about possible mental problems.

Yet, just as with the 2–4–6 task, there were an infinite number of possible explanations that would also be consistent with this set of evidence. He did not even consider one highly plausible alternative: that Bisma had an unusual disease that caused her to throw up, and she was sick of doctors who didn't understand her problem. To test this possibility, the doctor should have asked her questions like, "Do you think you are fat when others say you are thin?" and "Do you make yourself sick when you feel full?" Bisma would have gladly answered "no" to both questions, providing evidence that might make him less sure of his initial diagnosis.

Evian Water

Sometimes people purposely mislead us by the way they present confirming evidence, as in this advertisement for Evian water, which was circulated in the UK in 2004. The ad features a beautiful, naked woman, with parts of her body strategically obscured by a bicycle, proudly displaying her glowing skin. The copy at the bottom of the page reads: *Get skin so good you want to show it off. 79% of people who drink an extra litre of Evian pure natural mineral water a day notice their skin looking smoother, more hydrated, and as a result visibly younger.*

This sounds very convincing. But before you order a case of Evian water to prepare for your next trip to the beach, recall the 2–4–6 task. The rule turned out to be much broader than the hypotheses that subjects like Michael thought up; it wasn't a complex equation but rather any increasing number. Likewise, the truth behind the study results quoted in the ad could have been that drinking an extra liter of *any* water results in glowing, visibly younger-looking skin, whether it's Poland Spring, Fiji, Dasani, or even tap water—which would be much cheaper. Readers of the Evian ad who do not consider these other possibilities are victims of confirmation bias, which misleads them into thinking that only Evian could make them look younger.

Elevators

Here's another application of the 2–4–6 task that many readers must have personally experienced—the door-close button in elevators. When you're running late or simply impatient, you stab the button repeatedly until the door finally closes. Then, if you're like me, you take a deep breath, enjoying the satisfaction of knowing that you've saved yourself a few precious seconds of waiting. But how do you know that pressing the door-close button actually caused the doors to close? You might say you know it because every time you've ever pressed the button, the doors closed. But, as you also know, elevator doors close even when you don't press the door-close button, because they have timers. How do you know if the door closed because of the timer or because of you?

Thanks to the Americans with Disabilities Act, elevator doors are required to remain open long enough to allow anyone using crutches or a wheelchair to get in. According to Karen Penafiel, the executive director of the trade association National Elevator Industry, Inc., elevators' door-close buttons do not work until that waiting time is over. So, from now on, you can while away the seconds that you spend waiting for your elevator doors to close by patiently contemplating the pitfalls of confirmation bias.

Monster Spray

Many years ago, I took advantage of confirmation bias to comfort my son. When he was five years old, my husband became the Head (formerly called the "Master") of one of Yale University's residential colleges, which are just like Gryffindor or Slytherin in the Harry Potter books. We moved into the Head of Berkeley College house, which is a giant mansion built to house the Head and their family and also to host events for the students. The house is decorated in the typical Yale style, which is old, dark, gothic, and filled with portraits of people without smiles. Just imagine Hogwarts.

When Halloween arrived, the students decorated the house for one of the most anticipated parties of the year, turning it into a haunted house with spider webs, coffins, skulls, and similar goodies everywhere you looked. The decorations were so convincing that my son got very scared and wanted to move back to our former house. So I filled a spray bottle with water and told him that it was monster spray. We took it into every room of the house and he sprayed it. Ever since, not one monster has been spotted in the house.

"Bad" Blood

Confirmation bias can be collectively committed by whole communities for years and decades and centuries. The practice known as bloodletting is an often-cited case in point. From antiquity into the late nineteenth century, Western healers believed that if you drew out a patient's "bad" blood when they were ill, their ailments would get better. George Washington presumably died from this treatment when his doctor drew 1.7 liters of blood to treat a throat infection. Imagine two wine bottles filled with blood! How could our intelligent ancestors have believed for more than two thousand years that draining out a vital part of what keeps us going could be beneficial? By the time Washington was born, they had already figured out that the Earth is round and

Sir Isaac Newton had formulated the three physical laws of motion, but they still thought draining blood was the bomb.

Still, if we were in their situation, we might not have been much different. Picture yourself in the year 1850, with excruciating back pain. You've heard that in 1820, King George IV was bled 150 ounces and went on to live for another ten years. You've also heard that your neighbor's insomnia got cured by bloodletting. More importantly, you've heard that in general, about three-quarters of people who got sick and had blood drawn got better (I am making up these numbers for illustration). The data looks convincing. So, you try bloodletting and you actually do feel better.

But, here's the catch. Suppose there are one hundred people who got sick but did *not* have their blood drawn, and seventy-five of these people also got better. Now you can see that three-quarters of people get better *regardless* of whether their blood is drawn or not. It is possible to make a mistake like this because our bodies have the ability to heal themselves much of the time. Nonetheless, people neglected to check what would happen if they did *not* perform bloodletting. They focused only on the confirming evidence.

Quiz

When I give a lecture on confirmation bias, I grill students with the 2–4–6 task and all of the examples I have covered in this chapter so far, and then at the end of the lecture I give them a quiz. One of the questions I use, taken from

Keith Stanovich, Richard West, and Maggie Toplak's *The Rationality Quotient: Toward a Test of Rational Thinking,* may illustrate how hard it can be to discern confirmation bias.

> A researcher who is interested in the relationship between self-esteem and leadership qualities samples 1,000 individuals who have been identified as being high in leadership qualities. The researcher finds that 990 of these people have high self-esteem, whereas 10 have low self-esteem. Absent any other information, what is the best conclusion one can draw from these data?

> (a) There is a strong positive association between self-esteem and leadership qualities.
> (b) There is a strong negative association between self-esteem and leadership qualities.
> (c) There is no association between self-esteem and leadership qualities.
> (d) One cannot draw any conclusion from these data.

If you selected (a), you are like one-third of my students. That is incorrect.

I'm not telling you this to make fun of my students. I know for a fact that among the students who got this question wrong were so-called child geniuses, high school valedictorians, and national champions of math and debate competitions. They are also highly motivated to get this question right in pursuit of 4.0 GPAs. But confirmation bias can be powerfully misleading, even after they just learned about it.

As in the 2–4–6 task, the hypothesis that high self-esteem is associated with good leadership is a plausible first hypothesis, and on top of that, 99 percent of the data seems to support it. So, how could it be wrong? Once again, the problem is that the researcher in the scenario did not have any data on people with poor leadership qualities. If 99 percent of people with poor leadership qualities also have high self-esteem, then we cannot conclude that there is a positive association between leadership and self-esteem. Because the researcher did not have that data, the correct answer is (d). One cannot draw any conclusions from the data.

WHY IS CONFIRMATION BIAS BAD FOR YOU?

So far, confirmation bias may not sound detrimental to the person who commits it. The 2–4–6 task seems devious, intentionally designed to trick people, so those who don't solve this problem don't feel permanently dejected by their failure. The erroneous diagnosis of anorexia hurt Bisma, but not the doctor who made it thanks to his confirmation bias. Since we don't have scientists following us around in real life providing feedback on whether our conclusions are correct or incorrect, those of us who exhibit confirmation bias may not ever find out how many of our conclusions are wrong. Bisma's doctor probably still doesn't know how faulty his diagnosis of Bisma was—unless he reads this book. Given that those who commit confirmation bias may not even realize that their conclusion was wrong, can confirmation bias directly harm those who commit it? Definitely. It can hurt at the level of both individuals and societies.

Hurting Individuals

Let's talk about individuals first. Confirmation bias causes one to have an inaccurate view about oneself. Here is how that can go.

Many of us want to understand ourselves better and have an honest sense of where we stand in our lives and the world. We ask ourselves questions like: "Is my marriage in trouble?"; "Am I competent?"; "Am I likable?" We want definite, objective answers about our personalities, IQs, emotional intelligence, and "real" age. Our intense interest in ourselves explains the proliferation of all those "tests" on the internet and in magazines with titles like "What Does Your _____ Say About You?" (You can fill in the blank with *handwriting, laugh, favorite music, favorite food, favorite movie, favorite novel,* you name it.)

Imagine that a person named Fred notices an internet ad that asks, "Do you have social anxiety?" Fred is curious, so he pays $1.99 to take the test. When he finishes, he is told he scored high on social anxiety, being in the seventy-fourth percentile. Fred is initially skeptical, but now that he thinks about it, there have been times when he was socially anxious. He had trouble articulating his ideas at the last staff meeting, and he dreads going to cocktail parties. With all these confirming examples he just retrieved, he is now convinced that he has social anxiety. As in the 2–4–6 task, however, he forgot to retrieve disconfirming examples—like the staff meeting three weeks ago where he pointed out, without breaking a sweat, flaws in the current policy, or the fact that he enjoys talking to people as long as it's not during cocktail parties. Unfortunately, he has convinced himself that he is socially anxious, so he may avoid social situations even more than he did before. This is what's known as a self-fulfilling prophecy.

Here is another example of how confirmation bias can hurt those who commit it, and this time it's rather high-tech: DNA tests. It's easy to get a genetic profile these days through direct-to-consumer companies like 23andMe. You just need to pay about $100 to get your ancestry reports, and with another $100, you can find out about your health predispositions, like whether you're prone to type 2 diabetes or breast and ovarian cancer. According to one estimate, more than twenty-six million people in the United States had purchased direct-to-consumer genetic tests by early 2019.

But it's easy to misinterpret test results. Some people may believe genes determine our lives. Genes certainly don't do that, because they always interact with the environment.

Even when people don't necessarily believe genes are their fates, confirmation bias could make them rewrite their own histories as they attempt to make sense of themselves in light of their genetic test results. In a study I collaborated on with Matt Lebowitz, a former Ph.D. student of mine and now an assistant professor at Columbia University, we investigated that possibility.

We first recruited hundreds of volunteers who were willing to provide us with their mailing addresses so they could receive a package containing our experimental materials; they were told they would be compensated for taking part in our study. The packages they received included instructions for how to access our online experiment and a small plastic container with a label that read "Saliva Self-Testing Kit for 5-Hydroxyindoleacetic Acid," "Made in the U.S.A.," with an expiration date. After providing informed consent online, participants learned that as part of the study, they would take a saliva test that would measure their genetic predisposition for depression (participants were free to withdraw from the study at any point without losing compensation).

Then, they were instructed to open the plastic container and to take out a vial of mouthwash and a test strip from it. They were further asked to rinse their mouth using the mouthwash and spit it out. Unbeknownst to them, the mouthwash contained nothing but regular mouthwash with sugar mixed in by my research assistants. Then, they were told to put the test strip in the container under their tongue.

The instructions told them that the strip was sensitive to 5-Hydroxyindoleacetic acid, which serves as a proxy for detecting genetic susceptibility to major depression. The test strip was actually a glucose test strip, so when they put it under their tongues, its color changed in front of their eyes because of the sugar in the mouthwash they had just used. Participants clicked on the color they saw on the strip, and were told they would now learn what that color indicated.

At that point, our experimental program randomly assigned participants into one of two groups. One group was told that the color they entered indicated that they were not genetically susceptible to major depression. The other group was told that the color indicated that they were. Let's call them the gene-absent group and the gene-present group, respectively.

After the participants received the feedback, they took the Beck Depression Inventory II, commonly known as BDI-II. This is a well-validated measure of depression, which asks respondents about the levels of various depressive symptoms they've experienced in the past two weeks. For instance, for "sadness," they were given choices of, "I do not feel sad," "I feel sad," "I am sad all the time and I can't snap out of it," and "I am so sad or unhappy that I can't stand it."

We have no way of checking whether our participants' answers were accurate reflections of how their last two weeks had been. One thing we *can* tell is that given that the participants received one of the two genetic feedback conditions at random, there's no reason to expect that one group would have happened to have had a more depressing two weeks than the other. Some individual participants might have had worse weeks than others, but such variations should have been evened out due to the random assignment of a large number of participants.

Nonetheless, the gene-present group scored significantly higher on the BDI-II than the gene-absent group. That is, even though they were randomly assigned to receive one of the two genetic feedbacks, those who were told that they were genetically predisposed reported more depression from the previous two weeks than those who learned that they were not genetically predisposed. Furthermore, the average BDI-II score of the gene-absent group was 11.1, a score that is classified as having essentially no depression, while the average score of the gene-present group was 16.0, which is classified as experiencing depression.

Confirmation bias can readily explain this pseudo-depression. Upon learning that they were genetically susceptible to major depression, participants must have searched for times when they felt low in order to make sense of their "genetic test results." They might have recalled the night they could not fall asleep until 2 A.M., the morning they were not motivated to get to work, or that subway ride when they couldn't stop wondering about the meaning of their lives. All that confirming evidence must have made them believe that their past two weeks were more depressing than they really were.

Before moving on, I would like to be clear about this deceptive nature of the study, as I often receive questions about it. The experimental procedure was developed through extensive discussion with Yale University's Institutional Review Board, which oversees protection of human subjects. Upon completing the study, participants were told about the deception and the scientific value of the study and were provided with our contact information. To date, we haven't had any reports of adverse effects. One participant emailed us to ask for the brand of mouthwash we used, because she dislikes every mouthwash on the market but thought ours tasted really good; we had to remind her that that was because we added sugar to it.

Due to a mishap, we wound up with even more evidence of the power of the confirmation effect. Just after we began the study, we received a call from a police officer in Atlanta, Georgia, who told us that someone had brought in a suspicious package she received by mail, and they'd found our contact number in it. According to the officer, the woman who brought the package to the station had asked her family members whether any of them had ordered it and nobody claimed it. Interestingly, she also reported that when the package arrived, all her family members started feeling itchy! Because they believed the package might have contained something harmful like anthrax, they assumed their itchiness had been caused by the contents of the package. So, this was an example of confirmation bias operating in real life!

The woman who brought the package to the police officer lost only an hour or two of her day, and the family member who signed up for our study but then denied having done so lost the $10 that he or she would have received for their participation. But the type of confirmation bias the study revealed, as well as my earlier example of a personality test, illustrates a potentially much more profound peril of the confirmation bias, which is vicious cycles. That is what happens when you start out with a tentative hypothesis that becomes more certain and extreme as you exclusively accumulate confirming evidence, which in turn causes you to seek even more confirming evidence.

No genetic or personality test can provide definitive answers to who a person is. The results of these tests are always probabilistic. That is because the tests can be imperfect, but more importantly, it is also because that is the way the world is. For instance, the BRCA1 gene, which became famous thanks to Angelina Jolie's decision to undergo a double mastectomy when she tested positive for it, is considered to be one of the most informative genetic variants, predicting a 60 to 90 percent chance of developing breast cancer. Yet such high predictive power is extremely rare, because there are many, many nongenetic factors as well as multiple interacting genes that determine actual outcomes. Similarly, personality tests, which were developed to be used for workplace hiring and for counseling as well as to help us understand ourselves, provide highly decontextualized information; a person shown to be agreeable in a particular test might not be agreeable in a different environment or when performing another kind of task.

I'm not denying that these tests can be useful. I am planning to take a personalized genetic test soon to understand my health risks so I can be more proactive when it comes to aspects of my life I can control. In addition, understanding where I stand among the general population in terms of introversion-extroversion or open-mindedness can give me helpful insights into my social interactions.

Nonetheless, confirmation bias can easily lead us to a much more exaggerated and invalid view of ourselves. Once we start believing that we are depressed, we may act like a depressed person, making deeply pessimistic predictions about the future and avoiding any fun activities—which would make anybody feel depressed. The same goes with thoughts about one's competency: once you start doubting your competency, you may avoid risks that could have led to greater career opportunities, and then, no surprise, your career will end up looking like you lack competency. It can work in the opposite direction as well: a person may overestimate themselves, selectively remembering their accomplishments while ignoring their failures, and end up in an equally bad place. Because of such vicious cycles, I believe that confirmation bias is the worst of the cognitive biases that I am aware of.

As we will see next, these vicious cycles can also operate at the societal level.

Hurting the Society

We can begin with an episode that happened in my own family. When my daughter was in the first grade, my husband received the prestigious Troland Award from the National Academy of Sciences. The whole family went to Washington, D.C., for the ceremony. While waiting for the program to start, my husband was sitting on the stage with dozens of other awardees from various fields of science, and my two children and I were in the audience, along with some of the very best scientists in the United States.

At one point, my daughter asked me very loudly, "MOM, HOW COME THERE ARE MORE BOYS THAN GIRLS UP THERE?" Although dumbfounded, I was enormously proud of her for observing what she did. At the same time, I was embarrassed—not by my daughter's loud voice, but by the fact that I had not noticed that glaring display of gender imbalance on the stage myself. Being a scientist, I had probably become so accustomed to seeing more men than women in the field that I didn't really notice it anymore, although it was obvious to a child who didn't have any preconceived notions about our society.

I had no idea how to even begin to answer my seven-year-old daughter's question at the time, and fortunately for me, the ceremony started just a few moments later. But here is my belated answer: the reason that there were more men than women receiving awards isn't because "*only* men can be good at science." The truth is analogous to the "any increasing number" answer to the 2–4–6 problem: both men and women can be good at science. But our society has fallen for confirmation bias when it comes to men and science.

Traditionally, almost all scientists were men. Most people who are allowed to continue in their field do a good job. Thus, we developed the prevailing notion that men are good at science. Women were hardly given a chance to prove that they could be good scientists, too. Thus, we had little evidence that could disconfirm the belief that only men are good at science.

Because society believes that men are better at science than women, it continues to operate based on that assumption. When male students say something insightful during a seminar or in a class, they receive more compliments than female students who say similar things. Men are more likely to be hired and receive higher salaries than women who have identical credentials. Consequently, we end up with more eminent male scientists than eminent female scientists, which in turn strongly supports the notion that men are better at science than women. What is needed to rationally test this notion is to try to falsify this hypothesis by giving women fair chances. In terms of the reasoning fallacy committed, giving only men the opportunities and concluding that men are better is no different than a child believing that monster spray works because they sprayed it in every room and haven't seen any monsters since. We need to outgrow this fallacy.

How does this sort of confirmation bias hurt our society? Of course, it is a violation of the fundamental moral principle that all humans should be treated equally. And confirmation bias is irrational. But does it have negative consequences for our society that are even more tangible? Yes.

Here's a concrete example. Just now, I typed "scientists who developed COVID-19 vaccine" into a search engine to see how many female scientists would appear at the top. To avoid committing confirmation bias myself, I did not type in "*female* scientists who . . ." The first hit was Dr. Özlem Türeci, half of the power couple credited with developing the Pfizer-BioNTech COVID-19 vaccine. (The other half is her husband, Dr. Uğur Şahin, and they are both BioNTech's founders). Dr. Katalin Karikó also showed up as one of the scientists behind the Pfizer-BioNTech vaccine, and is mentioned as a highly promising contender for the Nobel Prize in chemistry. The fourth name I saw was also a female scientist, Dr. Kathrin Jansen, senior vice president and head of vaccine research and development at Pfizer Inc. What about the Moderna vaccine? According to Dr. Anthony Fauci, "That vaccine was actually developed in my institute's [National Institutes of Health's] vaccine research center by a team of scientists led by Dr. Barney Graham and his close colleague, Dr. Kizzmekia Corbett," a Black female scientist who also has been volunteering her time to help communities of color overcome vaccine hesitancy. These names were all on the first page of my search results. Only two males among them. Imagine what the world would be like right now had any of these female scientists been discouraged from pursuing her studies by her parents and teachers, if she had looked at the number of men receiving awards for their scientific accomplishments and assumed, like so many, that women couldn't do science.

It's not difficult to see how any stereotype based on race, age, sexual orientation, or socioeconomic background can work the same way. When only a few members of minority groups are given opportunities to show their capacities in certain fields, few of them obviously excel in those fields. That not only reflects poorly on society, but it robs us all of the advances that would be made by a broader pool of talent. A 2020 report from Citibank group quantified the ways that race-based discrimination and a lack of equal opportunity hurts America. Had our society invested equally in the education, housing, wages, and businesses of both white and Black Americans over the past twenty years, America would have been $16 trillion richer. In case that number is too large to grasp, the gross domestic product of the United States—the market value of all the finished goods and services produced within the United States—was $21.43 trillion in 2019. The $16 trillion was arrived at based on potential wages that Black workers could have earned had they gotten college degrees, the sales in the housing market that would have occurred had Black applicants received home loans, and the business revenue that would have been part of the economy had Black entrepreneurs received bank loans. We would have had $16 trillion to address climate change, fix health care, and work toward world peace, if there had only been no confirmation bias.

WHY DO WE HAVE CONFIRMATION BIAS?

If confirmation bias is so bad, then why do we continue to have it? How could it have survived throughout human evolution when it hurts us so much as individuals and societies? Does confirmation bias confer any benefits?

It may sound ironic, but confirmation bias is adaptive, helping us to survive, because it allows us to be "cognitive misers." We need to conserve our brain power or cognitive energy for things that are more urgent for survival than being logical. If some ancient ancestor found delicious berries in Forest X, why would he bother to see if Forest Y also has good berries when Forest X has been working so well? As long as he could get enough berries from Forest X, it didn't much matter whether only Forest X has good berries or all forests do.

Herbert Simon, who in 1978 was the first cognitive scientist to receive a Nobel Prize (it was in economics), made a similar point but did so as a more general principle, not just limited to confirmation bias. To understand his idea, first note that there are endless possibilities in the world. The number of possible chess games, even with its limited number of pieces and well-defined rules, is estimated to be 10^{123}, which is greater than the number of atoms in the observable world. Imagine how many possible versions of our future lie ahead of us. Thus, we need to stop our searches when they are satisfying enough. Simon called this "satisficing," a word he created by combining "satisfying" and "sacrificing."

Later studies by a variety of scientists found that individuals vary widely in how much they maximize or satisfice the kinds of searches they need to make throughout their lives. (For those who like to take personality tests, there are tests you can take online for free that measure where you stand on the scale of maximizer/satisficer.) Maximizers are always on the lookout for a better job, even when they are satisfied with their current one, fantasize about living a different kind of life, or write several drafts of even the simplest letters and emails. Satisficers are those who do not experience much difficulty shopping for a gift for a friend, easily settle for second best, or don't agree that one has to test a relationship before they are sure that they want to commit to it.

Interestingly, satisficers are happier than maximizers. That makes sense. Instead of obsessing over their search for a perfect soul mate, satisficers may settle down with someone good enough and enjoy the relationship. In the same way, enjoying good enough berries found in one forest makes people happier than trying to find out whether only that forest offers good berries or whether other forests offer berries that are as good or even better. Confirmation bias might be a side effect of meeting our need to satisfice, stopping our search when it's good enough in a world that has boundless choices. Doing that can make us happier and it can also be more adaptive. Nonetheless, the problem with confirmation bias is that we continue to use it even when it is maladaptive and gives us wrong answers, as we have seen through the many examples in this chapter.

COUNTERACTING CONFIRMATION BIAS

When we recognize the adaptive origin of confirmation bias, it also becomes clear how challenging it would be to eliminate it. In a later variation of the 2–4–6 task experiment conducted by other researchers in an attempt to eliminate confirmation bias, participants were plainly told how a number triplet can show them that their hypothesis is wrong, and that they can test their hypothesis with number triplets that they *don't* think will fit the rule. Even though it is quite an explicit instruction, it still did not help them find the correct rule. The strategy of trying to prove oneself incorrect must be super-confusing when the goal is to find a correct rule.

But because confirmation bias is so entrenched, we can exploit it to overcome it. This is not as paradoxical as it sounds. The key here is to consider not just one but two mutually exclusive hypotheses and try to confirm both. Let's consider a variation of the 2–4–6 task to see how it works.

Suppose I have two categories in my mind. Let's give them some arbitrary names so you can track them, say DAX and MED. Each category is defined in terms of rules regarding sequences of numbers. Your job is to find out what the rule is for each category.

To start, I can tell you that the sequence 2–4–6 belongs to DAX. You must figure out the rule for DAX and also the rule for MED by generating triples. Whenever you generate a sequence, I will tell you whether it's DAX or MED.

Michelle gives it a shot. Like most people, Michelle initially thinks that DAX is even numbers increasing by two, so she checks that first.

Michelle: 10, 12, 14.
Me: That is DAX.
Michelle (thinking, *Great, I think I figured out what DAX is, so MED might be odd numbers increasing by two. Let's check whether that's right*): 1, 3, 5.
Me: That is DAX.
Michelle: WHAT???

Note that while Michelle thought she knew what DAX is, she had to also find out what MED is. Accordingly, she generated a triple that she thought would be an example of MED. That is, she was looking for evidence that would confirm her hypothesis about MED. That sequence turned out to be a member of DAX, making her realize that her hypothesis about DAX was wrong. Let's continue watching.

Michelle (thinking, *Oh. So, DAX is then ANY number increasing by two. What could be MED then? OK, perhaps any number increasing by something other than two. Let's test that*): How about 11, 12, 13?

Me: That is DAX.

Once again, Michelle's attempt to confirm her hypothesis about MED turned out to be disconfirming evidence for her hypothesis about DAX. The rest follows the same pattern.

Michelle (thinking, *Fine, then DAX might be any number increasing by a constant. Then MED must be numbers that are NOT increasing by a constant. Let's check that*): OK, how about 1, 2, 5? (thinking, *That had better be MED.*)

Me: That is DAX.

Michelle (*Oh, so DAX must be any increasing number! Then MED must be non-increasing numbers. Let's check that*): 3, 2, 1.

Me: That is MED.

Michelle: Bingo! DAX is any increasing number and MED is any nonincreasing number.

Me: Correct.

Just like Michelle, 85 percent of the participants could solve the notorious 2–4–6 problem when it was framed as discovering two rules. This is what I meant by exploiting our tendency to confirm hypotheses to overcome confirmation bias. While people were trying to confirm their hypothesis about MED, they unintentionally disconfirmed their hypothesis about DAX. When a triplet that they thought was an example of MED or non-DAX turned out to be DAX, it revealed that their hypothesis about DAX was wrong, forcing them to revise it.

Now let's return to that dais at the National Academy of Sciences that my husband and all those mostly male scientists were on. We can apply the same strategy to overcome the confirmation bias that is the cause of the gender imbalance in the sciences. Suppose you start out with the observation that there are, say, fifty great scientists on the dais and they are all male. That leads you to think you understand what makes great scientists—Y chromosomes. Just as you need to figure out both DAX and MED, we now need to solve what makes bad scientists. Given your hypothesis about great scientists, you hypothesize that women make bad scientists. To verify that hypothesis, you give fifty smart women the opportunity to become scientists. Disconfirming your original hypotheses, they all turn out to be great scientists.

Here is another variation of the same strategy: ask a question framed in two opposite ways. For instance, in thinking through how happy you are with your social life, you can ask yourself whether you are happy or whether you are unhappy. These two questions inquire about the same thing, and should elicit the same response—like "I'm sort of happy"—no matter how the question is framed. Yet if you ask yourself whether you are *unhappy*, you are more likely to retrieve examples of unhappy thoughts, events, and behaviors. If you ask yourself whether you are *happy*, you are more likely to retrieve opposite examples. Indeed, participants in a study ended up rating themselves to be significantly unhappier when asked whether they were unhappy than those who were asked whether they were happy.

To avoid this sort of confirmation bias, we should query ourselves to generate evidence for both possibilities. And there are plenty of possible applications of this method. "Am I an introvert?"; "Am I an extrovert?"; "Am I bad at science?"; "Am I good at science?"; "Are dogs better than cats?"; "Are cats better than dogs?" Does the order of the questions matter? Yes, it does, as answers to the first question would likely bias the answers to the second question, and we will dive into that issue in a later chapter. For now, the point is that it's important to give both sides a fair chance.

Remaining Challenges

Making sure we figure out both MED and DAX and reversing a question's framing seem like straightforward-enough ways to mitigate the confirmation bias. Perhaps we could add those methods to the high school critical-thinking curriculum, and voila, the world would be a more rational place, right? Unfortunately, there are reasons why testing the alternative possibility—or figuring out what MED is—may not always be feasible.

Often, it's too risky. Think about the lucky underwear you always put on for your exams, or any other sort of ritual you perform in preparation for important meetings or matches. During his NBA career, Mike Bibby clipped his fingernails during timeouts. The Detroit Red Wings throw an octopus on the ice before each hockey game. Björn Borg grew a beard before each Wimbledon match, and only before Wimbledon. To prove that these rituals are not necessary, you would have to be willing to accept the risk of not doing them, thus foregoing their supposed protection. You'd have to show up for your exam wearing ordinary underwear, or play hockey without an octopus.

Similarly, the reason why bloodletting persisted so long is that if one has been taught that bloodletting actually works, then it would be unconscionable not to follow this "best practice." I myself swear by echinacea as a treatment for a cold, despite the mixed scientific evidence. I would not take the chance of not taking it the next time I feel under the weather. I know it is confirmation bias, but it is not worth overcoming it if doing so may cause me to sacrifice five days of my life to an illness that I believe I could have prevented. Likewise, if a person has been happily married for a long time, it would be absurd to run off with another person just to test whether there really is something special about your spouse.

As a more modern-day scientific example of a belief that people were reluctant to disconfirm, consider the "Mozart effect," prompted by a study published in the prestigious scientific journal *Nature* in 1993.

The researchers behind the Mozart effect reported that after listening to Mozart's Sonata for Two Pianos (for classical music fans it is K. 448), college students scored higher on a spatial reasoning test compared to those who hadn't listened to the music. The mass media took this finding to the next level, interpreting it as scientific evidence that babies who listen to Mozart develop higher IQs. Governors of states with some of the worst educational track records started handing out free Mozart CDs in maternity wards. Then, a "Baby Mozart" video was produced, showing colorful toys dancing to Mozart's music, and the company subsequently produced sequels in the "Baby Einstein" video series to encompass other geniuses, including *Baby Bach, Baby Shakespeare,* and *Baby van Gogh.* According to one estimate, a third of American households with babies owned at least one of those Baby Einstein videos around 2003. But as it turned out, the original Mozart effect was not longlasting, and was limited only to spatial reasoning rather than to the entire IQ ; some researchers could not even replicate the original finding. One study examined whether one of these bestselling videos help 12- to 18-month-old children learn new words better. The study found no difference at all between the children who watched the video for a month compared to the children who weren't exposed to the video and didn't receive any special training. Instead, the group who learned the words best were the children who learned the words that were featured on the video directly from their parents without using the video. Still, for parents who had a baby before this disconfirming evidence came out, buying Baby Einstein must have felt like a no-brainer. Even if they were not tiger parents, it would have seemed unconscionable to deprive their own precious babies of its potential good effects.

In addition to our reluctance to take a risk, confirmation bias is difficult to overcome because it is a habit. Just as we always start brushing our teeth on one side first without thinking, or we bite our nails, shake our legs, twist our hair, or crack our knuckles when we are nervous, we automatically and mindlessly confirm our hypotheses, as we do in the 2–4–6 task. Habits are difficult to break. For something like nail-biting, we could try wearing finger protectors or cutting our nails short. But where do we start if we want to break our habit of confirmation? Learning the devastating consequences of confirmation bias is the first step. And as another small step toward breaking the habit, perhaps you can begin a practice of disconfirming your assumptions about low-risk, everyday things by introducing some randomness into your life. In the same way that generating a random number sequence like "1, 12, 13" can inadvertently disconfirm one's hypothesis in the 2–4–6 task, you may accidentally find that what you have been enjoying or believing may not have to be your final answer. As it happens, there's even an app that can help you do this.

Max Hawkins, a computer scientist who used to work at Google, wondered what it would be like to be unpredictable. So he came up with an app that would randomly pick a location in his city from a Google listing and call an Uber to take him there without letting him know where the destination was. The first location his app took him to was a psychiatric emergency center, a place he had never imagined visiting. But this got him hooked on the exercise. He started discovering random florists, grocery stores, and bars he never knew existed because, thinking that his life was pretty well set, he'd never explored his options. He then expanded the app so that it would randomly choose public events within a geographic and temporal range that he specified and that were announced on Facebook, and he attended each one. He found himself drinking White Russians with Russian people, attending an AcroYoga class, and staying for five hours at a party thrown by a retired psychologist he had never met.

Although this sounds like fun when someone else is doing it, we may still be hesitant to buy his app because it seems daunting to commit to that much serendipity. So, here are some scaled-down exercises you can do to practice disconfirming. When you go to your favorite restaurant or order take-out, randomly pick an item from the menu; you might be surprised to discover a new favorite dish (or another least favorite dish, for that matter). Instead of following the same route on your way to work, try a new one. When going shopping with your friend, let your friend pick clothes for you so you don't buy yet another gray sweater or blue shirt. Eat lamb chop and salad for breakfast with a glass of milk, and cereal and an omelet for dinner with a glass of wine. Life is indeed full of possibilities, definitely more than the number of atoms in the observable as well as the unobservable world, and it's up to you to discover them.

3

THE CHALLENGE OF CAUSAL ATTRIBUTION

Why We Shouldn't Be So Sure When We Give Credit or Assign Blame

IN JANUARY 1919, AS THE WORLD was struggling to recover from the Great War and the 1918 flu pandemic, the leaders of the victorious powers met at the Paris Peace Conference to set the terms for the defeated countries. The negotiations soon came to an impasse, as Woodrow Wilson, the president of the United States at the time, did not want to punish Germany too harshly, while France and Britain demanded much harsher reparations. On April 3, Wilson came down with the flu, and he suffered from neurological symptoms even after he recovered. Although he was able to rejoin the peace conference, he didn't have the strength to pursue his agenda. As a result, the Treaty of Versailles included reparations that left Germany in massive debt. Many historians say the damage the treaty caused to Germany's economy paved the way to the rise of Adolf Hitler and the

Nazis. Accordingly, some have wondered whether we might conclude that if only Wilson had not caught the flu, there would have been no Holocaust.

Attributing such an atrocious and systemic crime against humanity as the Holocaust to Wilson's flu feels disturbing, even if the sequence of events is true. The causal explanation just doesn't feel right. Why is that? One possibility is that it's not such a good explanation. Even if Wilson had stayed perfectly healthy through 1919, there is no guarantee that the peace treaty would have been less punitive, that Germany's economy wouldn't have struggled for other reasons, or that Hitler wouldn't have eventually risen to power.

But for the sake of argument, let's suppose that someone invented a time machine and managed to stop the flu virus from infecting Wilson, that the treaty was less punitive than it was, and that this sufficed to keep the Nazis from gaining power. Even if it were possible to carry out such an experiment, we might still hesitate before we called Wilson's flu the sole cause of the Holocaust, because the Holocaust had any number of other possible causes. To begin with, if Hitler's parents had never met, Hitler would not have been born. If there were no antisemitism, then the Holocaust would not have happened either. What if Germany discovered that they were sitting on a gigantic oil field in 1919? What if Germany had won the First World War? Or if Archduke Franz Ferdinand hadn't been assassinated in Sarajevo and that war had never been fought? Although all of these and an infinite number of other possibilities might have thwarted the Holocaust, we don't blame Germany's lack of oil money, the assassination of the archduke, or the victory of the Allies for the Holocaust.

CUES THAT WE USE
TO INFER CAUSALITY

The number of possible causes for *any* event, not just historical ones, is infinite. Nevertheless, we can narrow them down to a smaller number of reasonable causes, and there is consensus on how to best do so because we use common cues or strategies in causal reasoning. That is not to say that we always agree on where to assign blame or credit. Some historians may argue that President Wilson's flu did cause the Holocaust. At the same time, we don't arbitrarily pick any cause. Few would say that a butterfly flapping its wings in Samoa in 1897 was the cause of World War II, but we would all agree that Hitler was one of them. We can agree about which causes are better and more plausible because we rely on similar cues when inferring causes of events.

In this chapter, I will describe some of the cues to causality that we commonly use. Here are samples of what I will cover in the chapter. Note that some lead us to blame Wilson's flu for the Holocaust, while others lead us to exonerate it. Our causal conclusions depend on which cues we rely on more heavily.

Similarity: We tend to treat causes and effects as
similar to each other. Perhaps we are reluctant to
attribute the Holocaust to Wilson's flu because
of their disproportionateness. Although Wilson
was an important man and the flu was a serious
disease, his illness was vastly different in scale
and severity from the systemic murder of six
million people.

Sufficiency and Necessity: We often think causes
are sufficient and also necessary for an effect to
occur. To the extent that Wilson's flu is thought
of as sufficient or necessary for the Versailles
Treaty to have turned out as it did, and the
treaty as sufficient or necessary for Hitler's rise,
we may think of Wilson's flu as the cause of the
Holocaust.

Recency: When there is a sequence of causal events,
we tend to assign more blame or credit to a more
recent event. Wilson's flu was temporally too far
removed from the Holocaust compared to more
immediate events, such as Hitler's rise, and thus
receives less blame.

Controllability: We are inclined to blame things
that we can control rather than things that we
cannot control. Wilson's flu was not something
that could have been reliably prevented, as they
didn't have flu vaccines back then, but one might
argue that Hitler could have been stopped before

he assumed power. Thus, we assign more blame to the latter event.

While going through these cues, it's important to remember that they are mere heuristics, or rules of thumb. In other words, while they can help us pick reasonable causes, they cannot guarantee that we will find the *true* cause. But they tend to give us reasonably good answers, so we rely heavily on them without realizing that they can also lead us astray. So next, I will discuss how an overreliance on any of them can cause us to draw the wrong conclusions.

Similarity

Imagine a yellow ball and a red ball on a pool table. If the yellow ball is moving toward the red ball quickly, then the red ball will move fast when the yellow ball hits it. If the yellow ball is moving slowly, the red ball would also move slowly. That is, the speed of the cause (the yellow ball) matches the speed of the effect (the red ball's motion). Similarly, loud sounds, like from an explosion, signal big impacts, while silence usually signals peace. Bad-smelling foods, like meat that expired weeks ago, tend to be bad for the body; good-smelling foods, like freshly picked strawberries, tend to be good for the body. In our daily life, causes generally match effects in terms of their magnitude or characteristics.

Since causes and effects tend to be similar to each other in reality, we pick up on that pattern and assume similarity when making causal attributions. That is, we are surprised if a cause is dissimilar to an effect. For instance, we expect large birds to make loud sounds, so if we hear a loud squawking sound and find out that it is coming from a tiny bird, we would be amazed and start recording it to share with all of our friends.

As another example, few people would believe that climate change—which affects biology, geology, economics, and essentially everything happening on Earth—could result from a single oil spill in the ocean; instead, most people understand correctly that it is caused by a myriad of human activities as well as natural disasters, interacting with Earth's atmosphere over time. Conversely, if the effect is simple, like broken glass on the floor, we assume that one person must have done it, rather than imagining that an entire family conspired to break a drinking glass.

Returning to the example that I used to open this chapter, the similarity heuristic is one reason we might feel that connecting Wilson's flu to the Holocaust is disturbing. Blaming the Holocaust on a single case of the flu seems to trivialize it. Even for those who are not fans of President Wilson, blaming his flu for the deaths of nearly six million Jews—along with the hundreds of thousands of gay men, Roma people, and disabled people who were also systematically murdered—feels far-fetched; instead, one grasps for a more palpably evil, state-level cause. This discomfort both illustrates and drives the similarity heuristic.

But relying on similarity to make causal inferences can

lead us astray, because causes and effects are not always similar to each other. While some good-smelling foods, like a
ripe strawberry, are good for us, a freshly baked cake made
of two sticks of butter and six eggs is not. And some bad-
smelling foods, like durian, kimchi, natto, and blue cheese,
are actually quite healthy. Though quiet typically signals
that there are no problems, a toddler's long silence can mean
trouble—the engrossed child might be testing how far a roll
of toilet paper pulls out, or exploring Mom's makeup drawer.

Folk medicine offers many examples of how an overreliance on similarity can be useless. Fox lungs were once
believed to be a cure for asthma, a lung disease. Rocky
Mountain oysters, which are deep-fried bull testicles, were
wrongly believed to keep men's testicles healthy and promote
the production of male hormones.

Conversely, because we rely on the similarity heuristic,
we may be reluctant to endorse even a certain cause when
it seems too dissimilar from the effect. For example, when
germ theory was first introduced to explain diseases, many
people were reluctant to believe it, because they could not
see how such tiny things as germs could be powerful enough
to harm or kill humans. That reluctance still lingers today.
During the pandemic of 2020, some people felt invincible,
refusing to wear masks and holding big parties in defiance
of all reasonable medical guidance. Had the COVID-19 virus
looked more like the White Walkers in *Game of Thrones* or
the zombies in *The Walking Dead*, public health management
would have been much easier.

The point of these examples is to remind you of the similarity heuristic's limitations. Sometimes small causes do produce large effects. For instance, we may believe that cheating a little may only do a little bit of harm, but cheating can have a cascade effect, affecting other people in unpredictable ways. Conversely, we may undervalue small acts of kindness, such as smiling at someone or asking them if they are OK. We should remember that seemingly insignificant gestures like those can make someone's day, or who knows, maybe even change their life.

Sufficiency

Although our causal judgments are influenced by similarity, it isn't one of the cues that we mainly use to figure out what caused a certain outcome. Sufficiency is a much more powerful cue.

Suppose Jill pours a bucket of ice water over Jack, and Jack screams. Philosophy Professor Phil comes out of his office and asks why Jack screamed. Jill confesses that it's because she poured ice water on him. Professor Phil is unconvinced. "How do you know that was the cause of Jack's scream?" he responds. (No, that is not a typical question one would ask in response to a situation like this, but Professor Phil specializes in epistemology, which is the study of how people know things.) Jill answers, "Because whenever someone pours ice water on someone, they scream." This is an example of a sufficient condition: *Whenever X happens, Y happens; when X is a sufficient condition of Y, we infer that X is a cause of Y.* So far so good.

The problem is that when we settle on one cause because it seems sufficient to result in the outcome, in many cases we are discounting other equally possible causes. Going back to Jack and Jill, once we learn that Jill poured ice water on him, we would not consider additional possible causes of his scream, such as whether there might have been a snake slithering toward him, or if Jack had just realized that he was late to his appointment with Professor Phil, and so on. That is, we think that the cause we have in mind is sufficient to result in the effect, so we ignore other causes that could also be responsible for it.

Discounting other potential causes works most of the time in the real world. Yet it is also important to be aware when we are doing this, because it can lead us to discredit people unfairly. Here's a concrete example. Suppose Gweyneth auditions for a TV show and gets the role. Michelle finds out that Gweyneth's father has a connection with the producer of the show. Michelle now believes that Gweyneth got the role because of her father's connection, and discounts the possibility that Gweyneth could be a good actor. But it is perfectly possible that Gweyneth has the connection and is also a great actor. We engage in this sort of discounting all the time. It's as if we believe that two causes are mutually exclusive, such that when one is present, the other is highly unlikely or couldn't have played a role.

We tend to think that if someone puts in a lot of effort to

succeed, that person is not talented. In my high school and college days, there were those annoying classmates who pretended that they hardly studied for exams, hoping it would make them look smarter. Allegedly, Mozart's widow burned 90 percent of his early sketches for his music to create the myth that he was such a genius that he composed everything in his head. Certainly, no one can deny Mozart's gift, whether his compositions came out of his head in their final states or not, but if the story is true, his widow was a shrewd publicist. As Michelangelo once said about his painting on the ceiling of the Sistine Chapel, "If you knew how much work went into it, you wouldn't call it genius."

Another well-known example of discounting is the relationship between intrinsic motivation and extrinsic reward. A person who enjoys cleaning his house may start to believe that he doesn't clean for pleasure once his father starts paying him for the task. In fact, a study showed that when people received a short-term monetary bonus, their performance improved, and when the bonus was removed, their productivity fell to a lower level than it was before. This probably happened because when they received the bonus, they attributed their productivity to it, discounting the intrinsic motivation they previously had. Thus, when the bonus was removed, they were left with lower intrinsic motivation than before.

Again, this sort of discounting is not necessarily fallacious because that seems to be the way things are in many real-life situations. If someone is not inherently interested in undertaking a task, you'll have to pay them to undertake it, so there is a negative relationship between intrinsic motivation and extrinsic rewards. Most of the time we don't get paid to do things we enjoy. I love walking my dog at dawn so I can watch the sunrise, but no one pays me to do it. It is also true that many talented people don't have to work as hard as less talented people to achieve the same outcomes. Nonetheless, focusing on a known cause and automatically discounting other equally valid causes can lead to any number of wrong conclusions.

Here is a real-life example of how such discounting can hurt others. In 2005, the economist, former U.S. secretary of the Treasury, and then Harvard president Larry Summers created quite a stir with his comments about the role of gender in science, which became one of the factors that led him to resign his position at the university. He stated that the gender gap in the sciences for people in high positions (like tenured professors) might be due to "issues of intrinsic aptitude, and particularly of the variability of aptitude." That is, even if the average levels are the same between men and women, he argued that there are more men than women who have the truly exceptional innate talents required for such high positions in sciences.

The subsequent controversy and debate in the academic community focused on whether there are indeed innate gender differences in science aptitude. Here, however, I want to discuss how the claim that there are differences in intrinsic aptitude is used to discount societal factors (like the society's expectations for girls and women) as the cause of the gender gap. According to *The Boston Globe*, "Summers said in an interview . . . 'Research in behavioral genetics is showing that things people previously attributed to socialization weren't due to socialization after all.'" Even if there are true genetic differences (I don't agree that there are, but I am supposing it here, just for the sake of discussion), those findings do not automatically rule out gender biases in socialization as the cause of the gender gap. Such unwarranted discounting has devastating real-life consequences, such as further widening the gender gap, as shown by a study inspired by Summers's comments.

The participants in the study were all women. First, they were given a passage to read disguised as a reading comprehension test, then they took a math test. The critical manipulation of the experiment was the contents of the passage. One group of participants read about a study showing that "males and females performed equally well on math tests." A second group of participants read a passage that "males perform five percentile points better on math tests than women because of some genes that are found on the Y chromosome." Reading that passage right before taking the test was sufficient to lower female participants' scores by about 25 percent! That's roughly the difference between a grade of A and C in my courses.

Critically, there was a third group of participants who were also told that males perform better on math tests than females, but the passage said that was because "teachers biased their expectations during early school formative years." That explanation was powerful enough to bring the participants' scores up to the same level as those who read that there was no gender difference in math scores. What this strongly suggests is that the participants in the second group, who learned about genetic differences between the genders, automatically discounted the idea that there could also be environmental differences. This remarkable study clearly shows that inappropriate discounting can harm performance.

Discounting a second potential cause for a phenomenon when one cause is known happens automatically. Sometimes, it can reflect the reality; but it can be blatantly wrong and damaging too, as we have seen. If we are cognizant of it, we will remember to be more careful before dismissing other causes and even try to prevent that by explicitly acknowledging the workings of other possible causes.

Necessity

Having sufficiently considered sufficiency in causal reasoning, let's look at the other side of the coin: necessity. A condition that is necessary for an outcome is a great candidate to be a cause of that outcome. In fact, it is a criterion used in legal systems, known as the "But for" rule.

Suppose Humpty Dumpty sat on a crumbling wall and had a great fall, breaking his skull. Further suppose the king, the owner of the wall, had been busy playing golf and had not ensured that his staff was maintaining the wall. If Humpty Dumpty's lawyer can establish that if not for (or "but for") the king's neglect, Humpty Dumpty would have been fine, then the king is liable for Humpty Dumpty's injury.

The "eggshell skull" rule, which originated from a case involving a thin-skulled plaintiff who died from a minor accident, further highlights the importance of necessity as a criterion to determine liability. Some legal scholars really do use Humpty Dumpty to illustrate it, so let's continue with that example. The king's defense lawyers could argue that Humpty Dumpty's injury occurred simply because his skull was so fragile—he's an egg, after all, or at least he was drawn that way, and eggs are notoriously breakable. But according to the eggshell skull rule, the king is still responsible, because even if the plaintiff had the preexisting medical condition of a fragile skull, the injury would not have occurred if the wall had been properly maintained.

We engage in similar counterfactual reasoning outside the courtroom, when we try to figure out what caused an outcome. Would B have occurred even if A hadn't happened? Would I have missed being involved in that accident if I hadn't gone to that store? Would they have stayed married if he hadn't taken that job? If the outcome would have been different in our counterfactual world, we treat that factor as a cause. There is nothing irrational about using counterfactual reasoning to make causal judgments; after all, it's what's used in the legal system.

Still, not all necessary conditions are causal. For example, oxygen is necessary for fire, but no one blames the presence of oxygen in California for its recurring wild fires. A person has to be born in order to die; if Marilyn Monroe had not been born, she would not have died. But the fact of Marilyn Monroe's birth has never been listed as a possible cause of her mysterious death. Before we can determine which of a plethora of necessary conditions is a cause, we need to complement the necessity heuristic with some other cues like the ones I will explain next. As a matter of fact, all of the cues I talk about supplement each other.

Abnormality

We tend to pick unusual events as causes. Being exposed to oxygen and being born are not atypical circumstances; there is oxygen in the air around us, and we all began our lives by being born. But the king's neglect of the wall that Humpty Dumpty sat on is an abnormal event, because the norm was that the king was supposed to take care of his properties, and thus his neglect is considered a cause. Similarly, intense back pain and the sound of a loud ambulance siren are each sufficient to raise anybody's stress level. But if you've been living with back pain for years, you would blame the siren for the stress you feel while it's blaring in your ear if it was a rare event to hear a siren. In contrast, if you lived next to a hospital but rarely had pains in your back, you would blame back pain as the reason for your stress.

This helps explain why people's causal attributions for the same event so often diverge; deciding what counts as normal or abnormal can vary depending on one's perspective. For example, let's say that Lin looked nervous during an interview. Lin is normally calm and confident, so from her perspective, it was the interviewer's grumpy style that was the cause. The interviewer, however, sees many other candidates, so the interview situation is normal for him. Lin seemed more nervous than the other candidates, so he marks it down to her personality.

Or consider gun violence. In the United States, people can legally purchase pistols, shotguns, rifles, and even semi-automatic weapons in some states. Whenever mass shootings occur, some people blame the shooters, reasoning that most gun owners don't go out and shoot people, so there must be something abnormal about those shooters, such as their mental health, anger management ability, ideology, etc. But from a global perspective, it is clearly the United States that is abnormal. According to the Small Arms Survey, the number of civilian firearms per 100 persons in the United States was 120.5 in 2018, the highest in the world. The United States has more than twice as many guns per 100 people than the runner-up, Yemen, and four times as many as Canada. Based on those statistics alone, there is clearly something abnormal about America when it comes to guns. Thus, from a global perspective the prevalence of guns in the United States is more responsible for the shootings than the individual character of the shooters.

People who are considering exactly the same event may draw different causal conclusions based on the perspectives they bring to it. When we wonder how someone could come up with a causal explanation that sounds so strange and nonsensical, it may be worthwhile to try to see the world through their eyes. They may still be wrong, but at least we'll understand how they reached their wrong conclusion. And who knows, we might even want to reconsider our own point of view.

Action

Another heuristic we use when picking a cause out of potential causal candidates is to blame actions more than inactions. Here's an adaptation of a classic example that illustrates this. Suppose Ayesha owns shares of Company A. She thinks about selling them and buying shares of Company B, and then does so. When Company B's stock takes a nosedive, Ayesha loses $10,000. Binita, on the other hand, has shares in Company B (the same one that Ayesha switched to). She had thought about switching to Company A, but decided to stay with Company B instead. Binita also loses $10,000, but it's not difficult to imagine that Ayesha, who had actively exchanged her shares in Company A for shares in Company B, would feel much worse than Binita, who did nothing.

There are numerous examples of blaming an action more than an inaction, even though the outcomes are identical.

If we learned that a foreign government kills twenty-five thousand innocent people every day for no reason, we'd be outraged, join a protest, write letters to our representatives, and search for ways to stop the killing. But if we read a UN report stating that twenty-five thousand people die every day from hunger and related causes (which is true), we feel sad, sigh, and shake our heads, but that might be the end of it—no protests, no writing letters to politicians. When someone deliberately kills someone, it's murder, and the punishment can be life in prison or death. But if someone watches a person dying whom they might have been able to save, they are considered guilty of negligent homicide, which carries a much lighter sentence—from six months to ten years in prison in most U.S. states.

We may place more blame on actions than inactions because when we are thinking about alternative possibilities, it's easier to think about one specific action we wish we hadn't done than to imagine all the things we might have done in cases where we did nothing. Had it not been for that evil government or that evil murderer, those people wouldn't have died. But inaction is harder to undo. In many cases, even if we did try to do something, it's not clear whether it could have averted the outcome.

Inaction is also invisible by definition, so we can easily neglect to consider how it could have caused specific events. Neglecting to combat racism or climate change, failing to report problems of equity that we witness, being complicit in maintaining the status quo, even when we know deep down there is a more just alternative—these are all examples of harmful inactions that might not be obvious to us.

Still, being oblivious to the costs of our inactions can cause problems that are irreversible. The fact that climate change may be unstoppable if we fail to take the right action now is just one example. Another is the consequence of not voting. People who don't vote may think it's harmless, but in doing so, they are taking away votes from a candidate who might have changed many people's lives had they won. Inaction is not always better than a bad action; sometimes it's equally bad.

Recency

When there is a sequence of events, we tend to credit the most recent one for the ultimate outcome. Whether in basketball, baseball, or soccer, teammates and fans celebrate the player who scored the last winning point in a tight game—like Michael Jordan's shot against the Utah Jazz that gave him his sixth championship with the Chicago Bulls. The losers, who could not stop the fate-changing play from happening, torture themselves by running it over and over again in their heads. Yet winning and losing is not solely determined by the last point, but by all the points accumulated throughout the entire game. Still, it's the last shot and the player who made it or missed it that get all the glory or blame.

You could argue that there is more pressure to score a

winning play or defend a lead in the final moments of a game, so the emphasis on it is justified. That may be true some of the time. But the following experiment shows that to most people, temporal order matters—even when it clearly shouldn't.

Suppose two people, Firth and Secondo, each toss a coin. If both get heads or both get tails, they each win $1,000. If the coins come up differently, neither wins. Firth decides to go first and tosses heads. Secondo goes next and, oh dear, he tosses tails. There goes $1,000.

Who gets the blame? Nearly everybody (92 percent) says Secondo. Who should feel more guilt, Firth or Secondo? Again, a large majority say Secondo. If I were Secondo, I would feel mortified. But if I were Firth, I would be livid and demand $500 from Secondo; that seems like appropriate compensation for all the trouble he caused. Yet blaming the loss on Secondo is preposterous. Firth could just as reasonably be blamed for not having tossed tails. Or better yet, neither should be blamed; the coin tosses are random—no one has the power to make the coins come up a certain way—and independent—the coin doesn't remember how it landed in the previous trial. But we tend to blame the more recent event, even in cases like this one, where temporal order should not matter at all.

Why is that? When there is a chain of events, say A causing B, B causing C, and C causing D, the final outcome D is not just caused by A, but by the entire sequence of A, B, and C. Thus, we may not want to give the full credit to cause A for outcome D, because if there had not been B or C, D would not have happened, even if A had. Yet if C happened, D would happen even if there had not been A or B. That is, C seems to deserve more causal credit than A or B.

The problem with the coin toss is that we tend to apply the same heuristic to cases where the sequence is of events that are not causally related. Firth's tossing heads did not cause Secondo to toss tails. Their coin tosses contributed equally and independently to the outcome. The last touchdown that upsets a tight football game is called the "winning touchdown," but the one before that was equally crucial to the win. When we give too much credit to the most recent event, even in situations in which the order of events should not matter, we are not only ignoring the other factors that are responsible for the outcome, but depriving them of their fair share of credit or blame.

Controllability

Before going over the last causal reasoning cue that I will explain in this chapter, let's step back and think about why we ask "why questions," as the answers will help us understand the basis of using the cue I'm about to talk about. Why do we constantly engage in causal reasoning? For example, when your date is late for dinner, why should it matter whether it was because his car broke down or because he almost forgot about your plans?

One of the most important functions of causal reasoning is to control future events. We want to avoid mishaps and repeat good outcomes by identifying the reasons each happened. If your date's car broke down, you might be more likely to want to continue the relationship than if you learned he wasn't looking forward to getting together nearly as much as you were. The cause of his delay helps you figure out if you want to dump him or not.

This leads us to an important and helpful cue we use: whether factors are controllable. Because we make causal attributions in order to guide our future actions, we typically don't blame things that we can't control. For example, when I burn my fingers lifting a lid off a hot pot, I can blame myself for not using an oven mitt and the next time I use that pot, I'll wear an oven mitt. In this case I don't blame burning my fingers on the fact that I have fingers or that heat transfers, because I cannot do anything about anatomy or physics. And while I might blame the pot manufacturer for selling a pot with handles that become scorching hot, I'm more likely to blame myself for buying that pot, as I can buy a new pot that has heat-resistant handles and I can't do anything about the manufacturer's decisions.

Our propensity to assign blame when we believe there were controllable elements can result in radically different emotional reactions to the same outcome. Suppose Steven was coming home from work and got stuck in a traffic jam caused by an accident. When he finally arrives, he discovers his wife has had a heart attack and it's too late to save her. Obviously, Steven would feel horrible about what happened. But he came home late because of a traffic jam, something that was beyond his control. He feels grief but not guilt.

Now consider a slightly different version of this scenario. Just as in the first one, Steven arrives home too late to save his wife, but in this case, it was because he stopped at a store to pick up some beer. Steven will likely blame himself for his wife's death for the rest of his life, thinking, "What if?" over and over again.

But blaming controllable actions can also lead to utterly tragic conclusions. Consider crime victims, many of whom blame themselves. A victim of the heinous crimes committed by Jeffrey Epstein was interviewed on NBC's *Today* show. She started giving him "massages" at the age of fourteen and then he raped her. The interviewer asked her, "In your mind, did you use the word rape? Did you recognize it then?" The woman answered, "No, I don't think I did. I just thought, like, you know, it's my fault."

Of course, there are numerous sociological and cultural explanations for why victims blame themselves. In terms of causal attributions, it happens because it is easier to imagine undoing their own behaviors than the perpetrators'. Survivors could be thinking, "If only I didn't have those extra drinks," or "What if I hadn't smiled at just that moment?" In their minds, those are all behaviors they might have controlled. The perpetrator's behaviors are much more difficult to change. Accordingly, the victims blame themselves, even though the perpetrators are obviously at fault.

OVERTHINKING AND RUMINATION

Causal reasoning can be very easy, as in the case of determining what caused Jack to scream when Jill poured cold water on him. But causal reasoning can also be more complex, as in the case of explaining why there are not enough female scientists. In some truly challenging cases, it may feel like we can't figure out the causes of a particular outcome at all, no matter how many cues we use. So as a final thought, let's consider cases in which causal questions are nearly unanswerable.

Probably, one of the most difficult why questions is "Why me?" When a string of bad things happens to someone, this question naturally pops up in their mind. This leads to rumination, continuously thinking the same thoughts, and still more why questions. *Why is this happening to me? Why can't I fit in? Why does this bother me? Why can't I move on?* When we keep trying to find answers to questions that are probably unanswerable, we may start to feel worse and worse.

Susan Nolen-Hoeksema, a colleague of mine at Yale University who passed away at the age of fifty-three, showed through her groundbreaking research in the field of clinical psychology how rumination can cause depression. In her study, undergraduate students were recruited based on their levels of depression. One group of participants was moderately dysphoric, meaning they had not necessarily been diagnosed with major depression but they displayed some depressive symptoms. The other group of participants was nondysphoric.

During the study, all the participants were instructed to think about their thoughts and emotions, such as "your current level of energy," "what your feelings might mean," and "why you react the way you do." Note that these are actually neutral questions, not intended to induce depressive thoughts. Participants engaged in this rumination task for eight minutes. Do not try this at home if you are dysphoric, because when their depression was measured again, those who were dysphoric had become significantly more depressed, merely by thinking about the reasons for their negative emotions.

Although rumination did not cause depression for the nondysphoric participants in this study, we might want to note that those who are generally happy can still be affected by rumination because we tend to ask more why questions when negative events occur and when we are in negative moods. We don't lose sleep trying to figure out why something went well, like passing a tough test or closing a successful deal. Instead, it's when things fail and we are in a dysphoric mood that we start getting obsessed with whys. As a matter of fact, people who live with chronic stressors, such as a loveless marriage, financial problems, or dissatisfying jobs, tend to ruminate more. The reason is plain: When people are dealing with their problems, they try to figure out their causes in an attempt to solve them and prevent future mistakes. They think they are gaining insights.

Unfortunately, studies also show that rumination actually prevents us from effectively solving our problems. This could be because of confirmation bias. When we feel low, we continuously regurgitate memories that confirm that feeling. It's hard to be a constructive problem-solver when you don't have faith in yourself. Because rumination doesn't help us discover solutions or causes, it can lead to further uncertainty, anxiety, and hopelessness about the future; it can spiral into alcohol abuse and eating disorders, as well.

One way to approach extremely difficult or nearly unsolvable causal questions constructively is to distance yourself from the situation. When we are ruminating, we tend to immerse ourselves in the problem. For example, when you are trying to figure out why a tragic event occurred, you may essentially relive that experience over and over. Obviously, this will bring up all the negative emotions again. When you are immersed in this way, it's also difficult to engage in problem-solving, as you become too emotionally depleted to maintain the necessary perspective.

Instead, it helps to distance yourself. Even if a problem affects only you, you can try to step back and take a different person's perspective on it. I'm quoting the instructions that were given to the participants in a different study, one that demonstrated the effectiveness of a self-distanced approach for resolving interpersonal conflicts. Participants were told to recall a time when they felt extremely angry and hostile toward someone. Then they were told: "Take a few steps back and move away from your experience . . . watch the conflict unfold as if it were happening all over again to the distant you." While they were maintaining this new perspective, they were told: "Try to think about the reasons underlying the emotions of this distant you." Compared to participants in the same study who were instructed to use the self-immersed approach, participants using this distanced perspective showed significantly less anger, both at a conscious and unconscious level.

Self-distancing also had a long-term benefit. Participants were brought back to the lab a week after the experiment in which they were told to distance themselves. In the second session, they were asked to reconsider the negative event. This time, they weren't told to self-distance. But even without the explicit instructions, participants reported significantly less negative emotions compared to another group of participants who hadn't distanced themselves from the negative experience during the first session. Once they'd seen the situation differently by distancing themselves, it was as if that new representation had stayed with them.

One big question remains, however, and that is: How can we tell when a why question is answerable or unanswerable? Strictly speaking, no why question is answerable. We can never find out the true causes of any outcome.

We can engage in counterfactual reasoning to consider whether the Holocaust might have happened had President Wilson not caught the flu, but we can never get a definitive answer of yes or no. We cannot change just one thing in the past and assume the rest would stay the same, as it never can (this is why I loathe most movies or TV series involving time travel, as it just couldn't possibly work the way the protagonists typically assume).

Even for apparently much simpler and less historical causal sequences, we can never be 100 percent certain what caused what. Suppose Sarah receives $100 for her birthday from her grandmother, and she becomes happy. But unbeknownst even to her, Sarah may have become happy because of the weather or a cute lizard she's just seen, or in anticipation of eating her birthday cake.

One might say that sometimes we can see causality in action. A red ball rolls over to a yellow ball, and upon contact, the yellow one starts moving. Didn't we just witness the red ball causing the yellow ball to move? Even when we are watching causal sequences with our own eyes, there is no guarantee that one event caused another, as noted by David Hume, a Scottish philosopher of the eighteenth century. The yellow ball may have been moved by some other force than the red ball, or even on its own. The perception of causality is an illusion.

When we believe we've found the right answer to a why question, in a sense all that we've really done is found the best answer to what we'd have to do if we want the same outcome to occur the next time we are faced with a similar situation—and what we should avoid doing if we want a different result. For that reason, perhaps the kinds of why questions that are worth trying to answer are those that potentially allow us to gain insights that can guide our future actions. If we are never going to encounter a similar situation, then it's not just impossible to pinpoint the answer, it's also pointless. Once you stop obsessing about why certain things happened, especially things you wish hadn't, then you can take a more distant view, which might help free you from negative emotions like remorse and regret, and also perhaps allow you to engage in more constructive problem-solving the next time you encounter a tricky situation.

4

THE PERILS OF EXAMPLES

What We Miss When We Rely on Anecdotes

I USE A LOT OF EXAMPLES IN MY teaching because cognitive psychology research tells me it's useful to do so. Vivid examples are more convincing, easier to understand, and harder to forget than decontextualized, abstract explanations. As an example (of course), consider the following:

If you need a large force to accomplish some purpose, but are prevented from applying such a force directly, many smaller forces applied from different directions may work just as well.

This is a highly abstract, decontextualized description, and while it makes sense, it's difficult to understand what situations it might be relevant to, so it's unlikely that anyone would remember it tomorrow. Now, consider the following story:

A small country fell under the iron rule of a dictator who ruled from a strong fortress. The fortress stood in the middle of the country, surrounded by farms and villages. Many roads radiated outward from the fortress like spokes on a wheel. A great general arose who raised a large army at the border and vowed to capture the fortress and free the country from the dictator. The general knew that if his whole army attacked the fortress at once, it could be captured. He gathered his troops at the head of one of the roads leading to the fortress, and prepared his attack. Just then, a spy brought the general a disturbing report. The ruthless dictator had planted mines on each of the roads. They were set in such a way that small bodies of men could pass over them safely, since the dictator needed to be able to move his own troops and workers to and from the fortress. But any large force would detonate the mines. This would not only kill many troops and render the road impassable, but cause the dictator to destroy many villages in retaliation. A full-scale direct attack on the fortress therefore appeared impossible. So, the general devised a simple plan. He divided his army into small groups and dispatched each to the head of a different road. When all was ready, he gave the signal. Each group marched down its road to the fortress, arriving at exactly the same time.

The vignette makes the same conceptual point as the abstract principle I cited previously, and while it's less concise, it's more engaging and memorable. Concrete examples are much more powerful than abstract descriptions, and they stick to our minds so much better.

They are also more convincing. In 1969, the U.S. Congress passed the Public Health Cigarette Smoking Act, requiring cigarette packages to carry a label reading "Warning: The Surgeon General Has Determined That Cigarette Smoking Is Dangerous to Your Health."The warning was so vague it had little effect. In 1984, Congress's Comprehensive Smoking Education Act was enacted, requiring specific warnings (e.g., that cigarettes cause lung cancer, heart disease, emphysema, pregnancy complications, and fetal injuries). But even these more specific warnings feel somewhat empty and bland. They don't make us gasp.

In Australia, the tobacco warnings are required to have photos next to them, like a picture of a premature baby with stick-like arms and oxygen cords attached to its wrinkled nose, or of repulsive green teeth alongside a warning about mouth and throat cancer. There is scientific evidence that disturbing images like these work. The CDC's antismoking campaign Tips from Former Smokers featured testimonials. One spoke through an electronic voice box after having a total laryngectomy as a result of throat cancer. Another displayed the mottled scars on his chest from his cardiovascular surgeries, and still another had had half her lower jaw removed because of oral cancer. The campaign increased quit attempts by 12 percent. In March 2020, the U.S. Food and Drug Administration finalized the rule that health warnings on cigarette packages should be accompanied by photorealistic images of smoking's negative health effects.

While vivid examples are a great way to communicate and convince, this chapter is about their perils. Specific examples and anecdotes can oftentimes be *too* powerful, leading us to violate important rational principles. In 2020, for example, it was not uncommon to hear people say things like, "My grandfather tested positive for COVID-19, and he recovered in one week. COVID is just the flu, after all," or "My friend never wears a mask, and he didn't catch COVID." For many people, one or two anecdotes from people they know are more persuasive than scientific evidence based on much larger samples.

For those who use social media like Instagram and Facebook, yes, we all know with our rational minds that our friends' magazine-grade pictures of fancy vacation spots, foods, and drinks are carefully curated moments of their lives and not the way they live day in and day out. But looking at that aquamarine-colored swimming pool, a Chanel bag placed next to a tropical drink, or the beaming faces of their friends, it is nearly impossible to imagine that they also have to deal with insecurities, anger issues, or occasional flare-ups of irritable bowel syndrome, just like everybody else.

In order to avoid being overly influenced by vivid examples and anecdotes, we may ask ourselves why they are so powerful. Some researchers have argued that it is because our minds are built to think in terms of what we experience and perceive rather than abstract concepts. That is, our thinking is based primarily on what we can see, touch, smell, taste, or hear. For instance, a picture of someone's mouth with oral cancer is convincing, as it makes you almost feel the kind of pain in your gum that you might have experienced at a dentist's office. While I agree with this proposal, it doesn't help us disregard a story we might have read in April 2021 about a mother of three who died of a blood clot after taking the Johnson & Johnson COVID vaccine. In our mind, that story can single-handedly override the CDC statistics that said that only six of the 6.8 million people who had received the J&J vaccine so far had developed such clots. So, let's reframe the question: Why are we swayed *less* by abstract statistics than examples of specific cases?

DATA SCIENCE 101

The main reason we're not convinced by statistics is that most of us do not fully understand them. There are at least three key concepts that all of us need to better understand if we are to avoid making blatantly irrational judgments in everyday life. They are: the law of large numbers; regression toward the mean; and Bayes' theorem. These terms sound technical, and some readers may be turned off just by seeing them named. Yet studies have shown that learning about these principles actually helps people make more accurate assessments. What follows is an explanation of each. And don't worry: I will use plenty of examples.

Law of Large Numbers

The law of large numbers is one of the most important principles to follow when we need to make inferences from limited observations. It simply means that more data is better. For example, we would be more confident in concluding that a new restaurant is excellent after having five great meals there than we would after having just one. The more observations we make, the more accurately we can generalize the pattern to cases we have not yet observed or make predictions about the future. Although we intuitively understand the law of large numbers, we frequently ignore it.

There are numerous real-life examples of ignoring the law of large numbers and going with an anecdote instead. I have already covered a few, but here are some more. The large majority of start-ups fail—somewhere between 70 and 90 percent of them, depending on whom you ask. Yet a story about how three guys went from renting mattresses to creating Airbnb, a $31 billion company as of 2020, can make anybody fantasize about becoming a rich entrepreneur.

Here's an example involving climate change. Despite numerous statistics showing the increasing levels of atmospheric carbon dioxide across thousands of years, climbing average temperatures, and rising sea levels, a single snowstorm can make the president of the United States tweet, "What the hell is going on with Global Waming [*sic*]?" Stephen Colbert had the perfect response: "Global warming isn't real because I was cold today! Also, great news: World hunger is over because I just ate."

It would not be prudent only to cite anecdotes to illustrate the problem with putting too much faith in examples, so let's talk about scientific evidence based on more controlled experiments over larger samples. One study used undergraduate students to gain insights into a topic that matters greatly to them: course evaluations. At the end of each term, most colleges ask students to rate various aspects of the courses they just took. One group of participants received the mean course evaluation ratings of former students, such as "Mean evaluation for the course (based on 112 of 119 enrolled): Good." The other group watched videos of a handful of students' verbal comments, such as, "I took the learning and memory course, and I rated it good ... It covers learning and memory fairly well, although being so general it can't go into as much depth as one might like ... At times I found myself bored ... But there was a substantial amount of worthwhile information." All participants—the ones who received only the ratings and the ones who only heard the anecdotes—then selected the courses they thought they'd want to take over the next few years. The results showed that the anecdotal comments had affected participants' choices much more than the mean course evaluation ratings, which were based on many more students' experiences.

In the spirit of the law of large numbers, I will present

another study, but this time, the researchers also tried to see whether participants could avoid being overly affected by a single case if they became aware of this reasoning fallacy, just as readers of this chapter have. Participants were told they would earn $5 for completing a survey that had nothing to do with the actual study. When they finished, they received their payment in cash, along with an envelope containing a letter of solicitation from Save the Children, describing a food crisis in southern Africa, which they were instructed to read carefully.

For one group of participants, the letter was comprised of factual information taken from the Save the Children website, such as: "Food shortages in Malawi are affecting more than three million children. Four million Angolans—one third of the population—have been forced to flee their homes." On average, those participants donated $1.17.

Another group was given no statistics. Instead they were shown a picture of a little girl named Rokia, a seven-year-old from Mali, accompanied by a story about the severe threat she faces from hunger and starvation. The average donation from this group was $2.83, more than twice as much. One way of interpreting these results is that participants were more convinced about the food crisis in southern Africa when presented with only a single case than with millions of cases. If so, this violates the law of large numbers.

The study also had a third group of participants who were taught about this identifiable victim effect. Would learning about its absurdity dispel it? The participants in what we call the "intervention condition" were divided into two groups: one read the data about the sufferings of millions and the other read the story about seven-year-old Rokia. But both groups also read the following text:

> Research shows that people typically react more strongly to specific people who have problems than to statistics about people with problems. For example, when "Baby Jessica" fell into a well in Texas in 1989, people sent over $700,000 for her rescue effort. Statistics—e.g., the thousands of children who will almost surely die in automobile accidents this coming year—seldom evoke such strong reactions.

This text did make a difference in one respect: the group that had just read about Rokia gave $1.36 each on average, less than they would have had they not read the explanatory text about the "Baby Jessica" effect. Unfortunately, reading that text did not increase the average donation among the participants who were only given the statistics and not the example. Learning about the power of specific examples may have made people somewhat more rational, but it did not help in terms of overall donations to Save the Children. To put it more abstractly, learning about the absurdity of the identifiable victim effect did not make people more influence-able by larger data. Which is why so many organizations like Save the Children feature stories as well as statistics on their websites and in their fundraising efforts, along with pictures of beautiful children, which might be the optimal approach.

But another study has shown that there *is* a way to help people seek out more data and give more credence to it, namely, teaching them about *why* the law of large numbers is rational. I could describe that study, but to make my explanation more vivid and memorable, I will use myself as an example.

When my son was five years old, I signed him up for a beginner skating class. He managed to stand on the ice and walk a few steps, but by the end of the third season that was still all he could do (yes, third season, not the third session). I also signed him up for soccer when he was seven. During one of his games, I noticed that whenever the ball flew toward him, he would run away from it. Based on those examples, it seemed clear to me that he was not into sports.

However, in keeping with the law of large numbers, we need to think of *all* sports, not just soccer and skating, but tennis, volleyball, baseball, basketball, surfing, curling, rowing, rock climbing, bobsledding, dressage, archery—you get the point. Suppose there are one hundred different kinds of sports in the world. Statisticians call this the "population," that is, the entire pool of things under consideration. I had observed only two samples from it, skating and soccer, but I'd made an inference about all of it. Generalizing based on such a small number of samples relative to the overall population is problematic. Suppose that out of a population of one hundred sports, my son was actually interested in sixty of them. Even if he could have enjoyed more than half of the variety of sports, it's not improbable that the two his mom happened to pick to introduce him to first were not his thing. After all, there were forty of them he wasn't going to like.

As for my son, it's a good thing his high school required all students to join a sport. He became a captain of its cross-country team and he still runs regularly. Perhaps he wasn't running away from the soccer ball but just liked running.

Regression Toward the Mean

The next statistical concept, regression toward the mean, is not an easy one to understand. I first learned about it when I was in graduate school, but to be honest, I don't think I understood it then. Having taught this concept for several decades, I think I've finally figured out how to explain it. A good way to start is with the phenomenon known as "the *Sports Illustrated* cover jinx," which is often used as an example of regression toward the mean.

Right after an individual or a team appears on the cover of *Sports Illustrated,* their performance will often begin to decline. For example, the August 31, 2015, issue of *Sports Illustrated* has a cover photo of Serena Williams, one of the top tennis players in the world, looking closely at the ball she'd just tossed in the air to serve. The headline reads, ALL EYES ON SERENA: THE SLAM. Inside, the article said, "Serena has a chance to win a calendar Grand Slam for the first time in her career. . . . This year, Serena defeated Maria Sharapova in the Australian Open final, Lucie Šafářová in the French

Open final and Garbiñe Muguruza in the Wimbledon final." But no sooner did the issue hit the newsstands than Serena lost to Italy's Roberta Vinci in the US Open, without even reaching the final.

On September 4, 2017, Tom Brady, who had achieved four Super Bowl MVP awards and two NFL MVP awards at the time, was featured on the cover of *Sports Illustrated*. He was still with the New England Patriots that year, and the cover was promoting the new season with the headline, THE PATRIOTS PROBLEM: CAN THE UNSTOPPABLE DYNASTY BE STOPPED? ANSWER: NO. It turned out this cover was wrong too. The Patriots lost to the Kansas City Chiefs 42–27 in the opening game of the season.

Those are only two examples, but I'm not ignoring the law of large numbers. Check Wikipedia for a long, long list of the teams and athletes who have experienced the *Sports Illustrated* cover jinx going all the way back to 1954, the year the magazine launched.

If the jinx is real, why does it happen? Perhaps those who become famous enough to be on the cover get arrogant and let their guards down. Or, they might become overly anxious due to the spotlight it shines on them. But rather than blaming the athletes themselves, we can explain the jinx as a statistical phenomenon known as regression toward the mean. What follows is an extreme example that has been made up to explain this concept. I'll come back to the jinx afterward.

Imagine ten thousand students take a 100-question true/

false judgment test. Let's say none of them have any background knowledge about the questions, which are along the lines of, "Jennifer Lopez's social security number ends with an even number," and "Ruth Bader Ginsburg had 15 pairs of sports socks in 2015." All of the students have to guess. In other words, there is no variance among the students' true abilities in answering these questions. But because they are true/false, the average score on this test is not zero, but more likely to be 50 out of 100, with many of the students scoring between 40 and 60. Yet, it is possible—although very rare—that the luckiest person in this group might guess 95 answers correctly, and that another with rotten luck might end up guessing only 5 right.

Now, suppose these same ten thousand students take a new true/false judgment test, also with 100 questions, and again they are purely guessing. What would happen to those who got 95 or 5 in the first test? It would be highly unlikely that the one who received 95 the first time would be that fortunate again. And that person who managed to dodge 95 correct answers is also unlikely to be that unfortunate a second time. Therefore, the scores of extremely lucky test-takers will tend to go down and the scores of extremely unlucky test-takers will tend to go up. This has nothing to do with the students' knowledge, motivation, or anxiety. It's a purely statistical phenomenon called regression toward the mean; extreme scores in the first test tend to move toward the mean in the second.

Regression toward the mean happens not just in cases

where test-takers are guessing. Whether people are taking tests, or performing in sports, music, or any other activity, there are always random factors that affect their performance giving a result that is better or worse than what they are capable of. Bearing this statistical phenomenon in mind, the *Sports Illustrated* cover jinx becomes easier to understand. Top athletes are also affected by random factors, like playing conditions, the strength of the competition on their schedules, quality of rest and eating, random bounces of the ball, variability in refereeing, and so on. When these random factors work in their favor, the athletes are more likely to show off their true talent or exceed it. That's when we say, *Wow, what got into her today!* Those who performed well enough to be featured on the cover of *Sports Illustrated* are likely to have had many random factors aligned in their favor for a notable stretch. But statistically, it can't last forever and it won't. No champions have perfect records. I am not saying that top performers are merely lucky, and that when their luck runs out, they'll regress down to the level of an average player. But when one is playing at an extremely high level of competition, even a little bad luck can mean a loss, and hence the jinx.

If we ignore regression toward the mean, we can make the kinds of inaccurate causal attributions that are known as the regression fallacy. For instance, we may assume that an athlete became too arrogant or lazy after becoming famous, when their loss was in fact due to regression toward the mean. The same thing can also happen in the opposite direction, causing us to give undue credit to people. For instance, let's say a teacher comes up with a new teaching method designed to motivate students, and tries it out on the students who scored the worst on the last exam. When the students' scores go up, the teacher claims that it was because her teaching method had motivated them to study. Yet this could also be regression toward the mean; those who performed the worst on the first exam probably had some random factors working against them, like having a bad day or having to answer questions in the one area that they happened not to have studied. The chance that all those unlucky factors would play out again in the subsequent exam is low. Unfortunately for the teacher, the students' scores may have gone up merely because of regression toward the mean.

The regression fallacy can happen in job interview situations, and this is where the power of specific examples, which this chapter is about, can be problematic. Many hiring decisions are made after face-to-face interviews or auditions. Those who have made the short list for the interview or audition have already passed a certain threshold, so there is not much variance among the candidates, meaning that random factors can be enough to shift the final hiring decisions. During an interview or audition, many things can go well or badly for the candidates and many of them are out of their control. The interviewer could be in a bad mood because of the morning news they listened to in their car on the way to work. I heard about a candidate who showed up with mismatched shoes because they happened to be lying next to each other when she was rushing out of the house; just imagine how self-conscious she must have been throughout the interview. Or imagine a candidate who happens to be wearing a blue shirt that is exactly the shade the interviewer adores, or that the audition piece a musician is asked to play just happens to be something she has been working on all year.

On top of all these random factors that can work for or against candidates, the inherent problem with interviews or auditions is that interviewers observe only a thin slice of the person's performance. Making hiring decisions based primarily on interviews is a violation of—now we can use a technical term!—the law of large numbers. But given that face-to-face interactions are vivid, salient, concrete, and memorable, interviewers think they are observing who the candidate truly is, rather than a biased portrayal of the person tinted by random factors. And this impression of a small sample of qualities on exhibit that particular day can make the decision-makers ignore the records that more accurately reflect the candidate's skills, demonstrated over many years. A person who looks amazing and brilliant during an interview may not be as awesome once they are hired. Given regression toward the mean, that is what we should, to some extent, expect. And a person who didn't perform brilliantly in an interview—for instance, the candidate who looked nervous due to her mismatched shoes—could turn out to be a big catch the company missed.

When I was on the job market looking to be hired as an

assistant professor, I had the opportunity to observe many approaches to interviewing as they were practiced by an assortment of psychology professors. At one university, the search committee chair asked me what "metaphysics" meant (because during my job talk I had said I would not talk about the metaphysics of causality), so I said something like, "How things actually are in the world, rather than how people think about them." The chair said, "WRONG." (I still don't know what was wrong about that or what got into him that day.) Of course, I didn't get that job; many years later, one of the faculty members who was present apologized to me about that chair.

If you are currently going through interviews, you might wish that interviewers would read this chapter, so you could just go in with your resumes and letters of recommendation from people who have known you for a long time. Wishing is not your only alternative; there is something proactive you can do to avoid being a victim of the regression fallacy committed by others, and that is to increase your sample size. Because there are always random factors in the world, if you apply for as many jobs as possible, those random factors are more likely to cancel each other out, increasing your chance to land a job where your true skills and experiences will be appreciated.

But how can we avoid committing the regression fallacy ourselves? What should interviewers do, for instance? If possible, the most straightforward method would be to evaluate candidates solely based on their resumes. That may sound outrageous, but I actually know of someone who practices it—the search committee chair who ultimately hired me at Yale University, who told me he didn't believe in interviews. To fill up the thirty minutes of our painfully long interview slot, I had to generate questions about my teaching philosophy and research plans myself and then answer them. I got the job, and I chose it over offers from other places that put me through the traditional two days of grilling, so I have no complaints.

Not doing any job interviews might not be a feasible option for hiring decisions that require us to see the candidate in action, however. Resumes and recommendation letters may feel too impersonal and vague; we may believe that we can make a much better decision if we can see the real person even for a brief moment. The problem is that once we do, it is difficult to not be overly affected by that one impression. At the same time, we know better. After all, few of us are ready to get married to someone after the first date. We just need to remind ourselves of the regression toward the mean, and not let ourselves get overly blown away by one stellar performance, or be too bothered by a candidate's choice of shoes. Just as we go on multiple dates with someone before committing to a marriage, we need to sample multiple observations of applicants following the law of large numbers. It takes more time and effort to observe them in different settings, but in the end, it might be cheaper and easier than hiring the wrong person.

Bayes' Theorem

The third important statistical principle that can help us become more rational is Bayes' theorem. Once again, let's start with an example.

Most adults in the United States who were born before the 1990s have photographic memories of the attacks on September 11, 2001. Videos were played over and over again on TV, showing the hole in the tower or dust billowing through the streets. Pictures of the ruins and stories about how people were rescued from them were in newspapers and magazines. Nearly three thousand people lost their lives. Americans were devastated.

Tragically, some of that rage was targeted at American Muslims who had nothing to do with extreme Islamic groups like al-Qaeda, which carried out the attack. Hate crimes against Muslims soared. Mosques were set on fire. Muslim women walking their kids in strollers were attacked by a woman screaming anti-Muslim obscenities. A man in St. Louis pointed a gun at a Muslim family, shouting, "They all should die!" In 2015, *The Washington Post* reported that "anti-Muslim hate crimes are still five times more common today than before 9/11."

The counterterrorism measures the U.S. government put into place immediately after 9/11 were also targeted at Muslims. Federal agents searched neighborhoods where Arab, Muslim, and South Asian families lived. Thousands of young men who did nothing illegal were arrested, detained, or "interviewed" only because of their ethnicity. Some were held for months under abusive conditions. There were many attempts to stop this sort of ethnic profiling, including the American Civil Liberties Union's 2004 report, which concluded that it is both inefficient and ineffective.

But why is it that ethnic profiling is ineffective? Some might defend Islamophobia, arguing that it's not practical to search everybody, and cite the fact that the attacks on 9/11 were carried out by Middle Eastern terrorists. Probabilistically speaking, however, ethnic profiling is not justified at all. To fully understand why that is, we need to understand some basic concepts in probability theory—specifically, Bayes' theorem.

Imagine there's this thing, and all that we know about this thing is that it's a koala. What is the likelihood that this thing is an animal given that it's a koala? That's easy. It's 100 percent. Next, consider the reverse. There's another thing, and all we know about this other thing is that it is an animal. What is the likelihood that this other thing is a koala given that it's an animal? Definitely not 100 percent.

Great! You've already demonstrated your understanding of what is known as conditional probability. As the name says, conditional probability is the probability that something, say A (animal), is true, given that or conditional on the fact that some other thing, say B (koala), is true. Now, we've established that the probability of A (animal) given B (koala) is not the same as the probability of B (koala) given A (animal).

The example here is clear-cut, and the logic applies to all conditional probabilities. But people frequently confuse the probability of A given B as being equal to the probability of B given A. A famous study that demonstrated this confusion involves how we should interpret results from mammography.

Suppose there is a woman with breast cancer. I will refer to having breast cancer as A. As we know, the likelihood that this woman would have a positive mammogram, one that shows a lump in her breasts, is quite high. Let's call the likelihood of having a positive mammogram B. That is, the probability of B (positive mammogram) given A (breast cancer) is high. But because of this, people also think that if a woman, ignorant of whether she has breast cancer or not, receives a positive mammogram (B), it means that she is highly likely to have breast cancer (A). That is, they think the probability of A given B is also high. But this is not the case. Just because the probability of a positive result (B) given breast cancer (A) is high, it does not mean that the probability of breast cancer (A) given a positive result (B) is equally high.

To compute the probability of A given B, or $P(A|B)$, from the probability of B given A, or $P(B|A)$, we need to use Bayes' theorem, discovered in the mid-eighteenth century by a famous statistician, philosopher, and Presbyterian minister named Thomas Bayes. There are many theories as to how Bayes got interested in probability theory, but my favorite is that he wanted to undermine the philosopher David Hume's argument against miracles. For those who are curious, I will get to that after I explain the formula.

Bayes' theorem is often used to update an existing theory or belief, A, given new data, B. For instance, after having watched three great movies featuring Tom Hanks, you may believe that all movies featuring Tom Hanks are amazing. Now you've seen a fourth one that was bad (sorry, Mr. Hanks, this is just hypothetical; I'm a huge fan of yours). Given this new evidence, you need to update your confidence in the belief that all Tom Hanks's movies are great. Bayes' theorem specifies a rational way to update the belief. No wonder it's pivotal in data science and machine learning. It is all about learning how confident one should be about a certain belief after one just observed new data.

The formula itself—busier than Einstein's $E = mc^2$—does look rather scary, and it is difficult to understand at an intuitive level. Readers who don't care for the formula itself can safely skip the next few paragraphs and resume with the one that starts with "OK" (but if you want to find out the Bayesian take on miracles, you'll have to stick with me through the math).

Bayes' theorem is:

$$P(A \mid B) = \frac{P(B \mid A) \times P(A)}{P(B \mid A) \times P(A) + P(B \mid not\text{-}A) \times P(not\text{-}A)}$$

where P(A) and P(B) mean base rates of A and B, like how often breast cancer occurs and how often we see positive mammograms. And not-A means absence of A, like not having breast cancer. So, P(B|not-A) means, for instance, the likelihood that one shows a positive mammogram even when one does not have breast cancer (as can happen, due to dense breasts). Applying the positive mammogram example to this theorem, even if the probability that women with breast cancer show positive mammograms, P(B|A), is very high, say 80 percent, and the probability that women without breast cancer show a positive mammogram, P(B|not-A), is very low, say 9.6 percent, the likelihood that women who test positive in mammography have breast cancer, P(A|B), is only 0.078 or 7.8 percent. This likelihood is surprisingly low, and it's because the base rate of breast cancer in the population, P(A), is 1 percent. Here's the equation with all the numbers plugged in.

$$\frac{0.8 \times 0.01}{0.8 \times 0.01 + 0.096 \times (1 - 0.01)} = 0.078$$

This is so low that those who test positive on a mammogram need additional testing, and it's also why there's controversy about whether annual mammography should be recommended.

In a study conducted in the early 1980s, participants (including practicing physicians) were provided with these numbers and asked to estimate the likelihood that a woman with a positive mammogram has breast cancer. Did the physicians give better estimates? Nope. Most people, including 95 out of 100 physicians, said the probability is about 75 to 80 percent. In order for that probability to be that high, the base rate of breast cancer, $P(A)$, had to be ridiculously high, say 30 percent. That is, only if breast cancer inflicts one third of middle-aged women, rather than 1 percent, we can say that a positive mammogram means 80 percent chance of having breast cancer. Because breast cancer is much rarer than that, the chance that a positive mammogram would detect actual breast cancer is less than 10 percent.

This last point takes us back to Hume versus Bayes. Hume questioned the validity of Jesus' resurrection, given that, outside the Bible, no dead people were ever resurrected in all of human history, and that only a few witnesses reported seeing Jesus after his crucifixion. Bayes didn't publish anything to rebut Hume's argument, but according to modern philosophers and mathematicians, here is how he could have done it using his own equation. If one believes that the probability of Jesus' resurrection, $P(A)$, is high, then the likelihood that Jesus was actually resurrected given that he had witnesses, $P(A|B)$, can be high, assuming that these witnesses are as reliable as the breast cancer mammography. In other words, claiming that Jesus' miracle really occurred does not violate the rational principles of probability theory. Of course, if a reasoner does not believe that Jesus was the Messiah such that the $P(A)$ is very low, Hume's argument is rationally correct.

OK, that was a rather long detour to prove why Islamophobia is irrational and discriminatory. We were talking about the fact that the 9/11 attacks were so vivid and salient that they became imprinted in our minds. As a result, people may believe that if there's terrorism, it was carried out by Muslims. That in itself is a fallacy because of the law of large numbers; the sample size is too small to conclude that all or even most terrorist activities are carried out by Muslims. But what makes it even worse in this case is that people also confuse the conditional probabilities. That is, based on the belief that "if there is terrorism, it's by Muslims," they flip it and believe that "if a person is Muslim, that person is a terrorist." This is as nonsensical as saying that "if something is a koala, it is an animal" also means "if something is an animal, it is a koala."

A determined debater might say that even though the two aren't the same, the likelihood that an animal is a koala is much higher than the likelihood that a nonanimal thing is a koala. So, the reasoning goes: The likelihood that any random Muslim is a terrorist should be higher than the likelihood that any random non-Muslim is a terrorist, and therefore, ethnic profiling is statistically justified. Right? No.

As of 2021, the adult population of the United States is about two hundred million, of which about 1.1 percent, or 2.2 million, are Muslim. In my analysis here, I use a 2017 report by the U.S. Government Accountability Office, which reports the number of fatal terrorist incidents that took place immediately after 9/11 up to the end of 2016, which is the most recent record I could find. According to this report, violent extremists carried out eighty-five actions in the United States that resulted in fatalities between September 12, 2001, and December 31, 2016. Twenty-three of these incidents, or 27 percent of them, were attributed to radical Islamists. Of these, six were committed by the same person, the Washington, D.C., Beltway sniper in 2002, and three by the brothers who committed the Boston Marathon bombing. So, the total number of unique terrorists who were motivated by radical Islamist views and caused fatalities in the United States during that period is fewer than twenty-three; I count sixteen from the report.

Perhaps this number may seem shockingly low to some readers, who also distinctly remember the Orlando night club shooting and the office party shooting spree in San Bernardino, California. All these are counted in. (If you still feel like there has to be more, that is yet another effect of vivid examples, known as the availability heuristic as named by psychologists Daniel Kahneman and Amos Tversky: we judge the frequency of events based on how available they are in our minds.)

Now we are ready to calculate the likelihood that a random Muslim adult on the streets of the United States is a terrorist. That would be the number of Muslim terrorists, sixteen, divided by the total number of Muslims in the United States, 2.2 million, which is 0.0000073 or 0.00073 percent. That is, even if FBI agents detain ten thousand adult Muslims, the chance that one of them is a terrorist is nearly zero. (In case any readers remain skeptical about my estimate of sixteen Muslim terrorists causing those twenty-three fatal attacks, it shouldn't be difficult to see that even if that number is increased to 160, the probability is still essentially zero.)

The people who attempted to justify the ethnic profiling and discrimination against Muslims had no understanding of conditional probabilities. The likelihood that a terrorist on U.S. soil during the fifteen years sampled was Muslim is 27 percent. That is, if they had checked 100 known terrorists, 27 of them may be Muslims. That is substantially high, but that's not the probability to use when we make decisions about detaining people. It is the inverse probability that should be used, and that probability is essentially zero.

The image of the Twin Towers in flames and Osama bin Laden's face are scorched in our minds. Mix them with our confusion about conditional probabilities, and we fall into utterly outrageous prejudice, which harms innocent people.

MAKING THE BEST USE OF SPECIFIC EXAMPLES

Statistical reasoning is hard, and there are good reasons for that. We rarely work with large numbers or interact with the entire population from which samples are drawn. It is difficult to imagine all the random factors that underlie peak or terrible performance, causing regression toward the mean. The notion of probabilities was not even introduced to human culture until the 1560s. Even if the three statistical concepts that I have covered in this chapter are learnable, it is not easy to keep them in mind in everyday reasoning. I have been teaching these concepts for decades, but I often catch myself being overly influenced by anecdotes. Given that specific examples are so powerful, let's end the chapter with some ideas about how we can make the best use of them.

We may think that once we learn something through a powerful example, we should be able to apply it in new situations. After all, the whole point of learning is to transfer our knowledge to the new problems we will face in the future. Ironically, however, learning through examples comes with one important caveat. To illustrate it, see whether you can solve the following problem:

Suppose you are a doctor, and your patient has a malignant tumor in his stomach. It is impossible to operate, but unless the tumor is destroyed, the patient will die. A certain kind of X-ray treatment provides some hope. If the X-rays are beamed at the tumor in such a way that they hit it all at once and at a sufficiently high intensity, the tumor will be destroyed. Unfortunately, the healthy tissue those high-intensity rays pass through on the way to the tumor will also be destroyed. At lower intensities the rays are harmless to healthy tissue, but they will not affect the tumor. What type of procedure might

you use to destroy the tumor while at the same time
preserving the healthy tissue?

If you can't solve it, no worries. It's a hard problem, and
it is not an intelligence test. So here is a hint. Think of an
example that I have already presented, the story about the
general and the dictator's fortress that opened this chapter.
Now the solution should be straightforward. It is to beam
radiation from multiple directions to converge on the tumor.

In a study that used these two problems, college students
from the University of Michigan—which is to say, very
smart students—were first presented with three stories, one
of them the fortress story. To make sure they were not skim-
ming, they were asked to summarize them from memory.
Just four minutes later, they were presented with the above
tumor problem and only 20 percent of the participants
could solve it. Eight out of ten of those bright students
failed to remember and apply the example they had read
and summarized only a few minutes earlier. It likely took
you more than four minutes to read through this chapter;
it's no wonder if you didn't make the connection.

But if participants were provided with an explicit hint to
apply one of the previously presented stories, almost everybody could come up with the solution. This means that the
difficulty does not lie in applying a known solution to a new
problem, but in spontaneously retrieving it from memory.
That is bad news, because it means that four minutes after
a teacher explains a method to students, the students won't

be able to apply it to a new situation without an explicit reminder from someone else to do so.

But wasn't this chapter about how powerful examples are? If so, then how could students fail to retrieve them? There is nothing contradictory about this. Examples are so powerful that people are more likely to recall irrelevant details from them, like the fact that there was a general and a fortress, than the abstract principle of convergence that underlies that particular story.

Having identified this challenge, the researchers tried out various methods to help students spontaneously recover the underlying principles they learned from examples. The method that worked best was to demonstrate the same principle across multiple stories. For instance, you just learned about the convergence solution in the context of a general conquering a fortress and also in the context of a doctor treating a tumor. If you were confronted with a third problem requiring the convergence solution, you would be that much likelier to transfer those examples to it.

In other words, if you are telling a story to make a point, your point will have a greater likelihood of being remembered if you embed it in multiple stories and tell all of them. We talked about Jesus earlier. A master storyteller, he seems to have known this technique. To explain that God welcomes lost souls, Jesus told the parable in which the shepherd rejoices at finding one lost sheep even though he had ninety-nine other sheep who had not gone astray. Then he followed it with another parable, in which a woman searches high and low for one lost silver coin and celebrates when she finds it, even though she had nine other coins.

You may have noticed that I use not just one example but at least two to cover the same concept. Hopefully, you are more likely to spontaneously recall the law of large numbers the next time you see a group of kids playing soccer or receive a solicitation letter from a charity in your inbox. Seeing *Sports Illustrated* in the magazine section of a drug store, or having a too-good-to-be-true first date or first meeting with someone may remind you of regression toward the mean. The point about $P(A|B)$ not being the same as $P(B|A)$ may hopefully be brought back to your mind when you hear about non-Islamist terrorism, or encounter an animal that is not a koala.

5

NEGATIVITY BIAS

How Our Fear of Loss
Can Lead Us Astray

I ONCE WASTED A LOT OF TIME looking for a new phone case. The one I was using at the time had a picture of Snoopy; a bit too cute for a professor. I searched and searched a myriad of online stores. Remember the maximizer/satisficer scale I wrote about in chapter 2, the one that measures the differences in our tendencies to maximize searches? I received the highest possible score. When it comes to shopping, I can't stop searching until I find the perfect item. Finally, I came across a case that seemed quite promising. I liked the way it looked in the photograph and the reviewers' ratings were good, giving it an average of four out of five stars.

Then I started reading the reviews. The first four gave it five stars: "I love it! Great material and great look." "My boyfriend loved this case! Sturdy and easy to hold." "Excellent quality . . . perfect in every way . . . beautiful!!" "Sleek look, been 4 weeks, so far so good!"

Then, I saw a one-star review. "Very nice-looking case,

however, it is very fragile and uncomfortable to hold with one hand. The case broke within a week." The four positive, five-star reviews that I read earlier could not undo the damage that this single negative one did. What bothered me the most was that the reviewer reported that it broke within a week, even though the positive reviews specifically said the case was sturdy and still good after four weeks of use. So, I was stuck with Snoopy for another year.

EXAMPLES OF NEGATIVITY BIAS

You don't have to be a super-maximizer like me to be overly affected by negative information. In one study, researchers tested the ways that positive and negative reviews affect sales of electronic products, such as cameras, televisions, and video games. The researchers selected more than three hundred products introduced at Amazon.com between August 2007 and April 2008, collected their sales ranks and the percentages of positive (those with four or five stars) and negative (those with one or two stars) reviews they received, and then examined the relationships between them. As you would expect, the percentage of negative reviews was negatively related to sales rank, while the percentage of positive reviews was positively related to sales rank. But more importantly, the researchers compared the magnitude of the influences. The percentage of negative reviews had a much greater impact on sales ranks than the percentage of positive reviews.

Numerous psychological studies have shown that people weigh negative information more heavily than positive information, and not just when forming judgments about products, but about people. Suppose there is a man named John, and you've only seen him twice. The first time you saw him, he was eating with some friends in a restaurant. He didn't look particularly friendly or lively, but he seemed reasonably sociable. The second time, you were standing near a table set up outdoors under a poster reading SAVE OUR LOCAL BUSINESSES. John walked by without stopping, completely ignoring the woman who asked him to sign the petition. You might think that the somewhat positive behavior and the somewhat negative behavior you witnessed would cancel each other out, leaving you with a more-or-less neutral impression. But, people give more weight to negative behavior, so your overall impression of John is likely to be more negative than neutral.

Negative events also affect our lives more than positive events. Just one episode of childhood trauma, such as sexual abuse, can have harmful consequences that last a lifetime, including depression, relationship problems, and sexual dysfunction. Such episodes are not easily offset by other positive aspects of childhood, even if there were many more happy events than bad ones.

The negativity bias can affect us so severely that it can

cause us to make decisions that are blatantly irrational. For instance, we tend to avoid an option framed in terms of negative attributes, while we would gladly accept the exact same option if it is framed in terms of its positive attributes. Thus, people prefer flights that are on time 88 percent of the time over flights that are late 12 percent of the time. They judge a 95 percent effective condom to be better than a condom with a 5 percent failure rate. They would rather receive a 5 percent raise when inflation is 12 percent over a 7 percent cut when inflation is zero.

One of my favorite studies along this line involves ground beef. Twenty-five percent fat sounds pretty bad; it explicitly says that a quarter of what you're looking at is pure fat. On the other hand, 75 percent lean, which means exactly the same thing, sounds vastly healthier and better. We might think that we wouldn't fall for it if someone, as a marketing ploy, tried to make us think there was a difference. But that's not the case. In one study, researchers cooked ground beef and asked participants to taste it. The researchers don't report whether the beef was well-done or medium, or if they added any salt and pepper, but they do let us reveal one essential thing: the same ground beef was cooked the same way for everyone. The only difference was in how it was labeled: half the participants were told it was "75 percent lean," and the other half were told it had "25 percent fat." That made all the difference. People who tasted "75 percent lean" ground beef rated their hamburgers to be less greasy, more lean, and better in quality and taste than the people who'd eaten ground beef that was "25 percent fat."

Which Grades are Better, As and Cs, or All Bs?

I became interested in the negativity bias in the context of college admission processes and ended up conducting a study on it. I began the study around the time my older child started thinking about college applications. I purchased and read three books on college admissions, as I didn't know the applicants' side of the process, having attended college in Korea. In addition to explaining the technical aspects of admissions, every one of these books underscored the importance of students showing passion and enthusiasm in one specific area. One of them called it a "hook."

I have also noticed the similar emphasis while I sat on the other side of the process. At Yale, admissions committee meetings are run by highly competent professional admissions officers, but they invite one or two professors to join each meeting. Over the years, I had attended a few such meetings after attending a training session. That is where I saw the formal statement of Yale's admissions policy, created by Kingman Brewster, a former president of Yale. Written in 1967, the policy is still in use. It states, "We want as many [of our graduates] as possible to become truly outstanding in whatever they undertake. It may be in the art and science of directing the business or public life of the country, or it may be in the effort to improve the quality of the nation's life by the practice of one of the professions. . . . The candidate is likely to be a leader in whatever he ends up doing." In other words, successful candidates don't have to be perfect in everything—and they can do whatever they want—but they should be outstanding in one field. In fact, as the college admission guidebooks noted, this philosophy isn't limited to Yale. A *Washington Post* article nicely summarizes, "Colleges want a kid who is devoted to—and excels at—something. The word they most often use is passion." An article in *U.S. News & World Report* also listed passion as the number one way to bolster one's chances for college admission.

But it occurred to me that this emphasis on passion seems to contradict the robust phenomenon in psychology I've just been discussing: that people are more influenced by negative than positive information. To illustrate the discrepancy with a simplified example, let's say there are two rising seniors in the same high school, Carl and Bob. Carl has been receiving As and A+s in certain subjects, but also Cs and C-s in others. This pattern suggests that Carl is more passionate and enthusiastic about some subjects than others. Bob has been getting more uniform grades, receiving Bs, B+s, or B-s in all of his subjects. That is, he had no Cs but no As either. Assume that the GPAs of these two students are the same. If you are a college admissions officer and their transcripts are all the information you have, which student would you favor?

If passion is the essential feature, the admissions officer should favor Carl. But people tend to weigh negative
information more than positive information. If this negativity bias were to dominate the decision, then the A that
Carl received in, say, Chemistry, might not balance out the
damage caused by the C that he received in, say, English,
and the officer would prefer Bob. To find out the power of
the negativity bias in a situation where there were criteria in
place that should have counteracted it, I decided to conduct
an experiment.

First, we created transcripts for students like Carl and
Bob. To avoid any course biases, we created several, such that
the As and Cs were associated with different courses. Then
we recruited participants, who were tasked with choosing
which student to admit. Some were recruited from an online
platform; some were undergraduates who had, of course, recently been through the application process themselves. Finally, we recruited admissions officers from various colleges
and universities across the United States. When we asked
them to choose between a student with varying grades and
a student with homogeneous grades, the majority of participants chose the latter, the one without Cs, but also without
As. In particular, nearly 80 percent of the admissions officers preferred the student with more uniform grades.

Participants also judged that the student with uniform
grades was more likely to have a higher GPA in college,
and be harder-working, more responsible, and more self-
disciplined than the student with a mixture of As and Cs. In

addition, they predicted that the student with homogeneous grades would be more likely than the student with mixed grades to become an owner of a midsize to large business, have a career in executive management, or become a government official, lawyer, doctor, or engineer. The students with homogeneous grades were also predicted to earn an annual income higher than the student with heterogenous grades would. All this despite their equal GPAs and the schools' preference for passion.

We tried other variations of the transcripts to make sure the effect was replicable. In particular, passion and enthusiasm are mostly stressed as criteria for admission by highly competitive colleges, which expect much higher GPAs than what we used in the first study. So we did the experiment again, this time recruiting admissions officers only from the most competitive colleges—ones that all readers would certainly recognize as such. In addition, both hypothetical students had spectacularly high grades, with overall GPAs of 4.0 out of 4.3. This time, the student with homogeneous grades had all As except for one A+ and one A-. That is, there weren't that many A+s but the lowest grade this student received was an A-. The student with more heterogeneous grades had many more A+s, *eight* of them. But alas, that student also received three B+s. Still, the two students' GPAs were identical. Nonetheless, the negativity bias prevailed. Admissions officer participants preferred the A student who did not have any B+s to the A student who had B+s, even though the latter student had eight A+s.

Before I move on, I must put a big disclaimer here. Students should still put extra effort into their favorite subjects, and dedicate themselves to their passions. And students with uneven grades should not feel discouraged—plenty of such applicants will gain admission to a favored school. Remember that colleges review a lot more information than just GPAs, especially recommendation letters, extracurricular activities, and essays.

Loss Aversion

Given that the negativity bias affects so many different kinds of judgments, it should not surprise anyone that it also affects us when we make decisions involving money. Yet the specific ways in which the bias operates can be obscure.

During the 1970s, a field called behavioral economics began receiving a great deal of attention. Behavioral economics can be thought of as the intersection of psychology and economics; its main agenda is to research how human judgments and choices operate against the rational principles developed in economics.

Behavioral economics has revealed numerous cognitive biases and thinking traps, challenging economics' fundamental premise that people's behavior is based on logical choices. (Readers might have seen articles and internet postings with titles like "61 Cognitive Biases That Screw Up Everything We Do" or "Cognitive Bias Cheat Sheet: Because Thinking Is Hard.")

In 1979, Daniel Kahneman and Amos Tversky published one of the most important papers in behavioral economics, titled "Prospect Theory: An Analysis of Decision Under Risk." To quantify how much impact an article has, one index frequently used in academia is the number of times the article has been cited in other published articles. According to one index of citations that goes up to year 2021, this paper has been cited more than seventy thousand times. To understand how astronomically high that is, compare it to Stephen Hawking's 1973 paper on black holes, which has received about one-fifth that many.

One of the revolutionary ideas Tversky and Kahneman presented is the insight that we treat the same monetary values differently depending on whether they are gains or losses, which leads to what is known as loss aversion. Many readers may have heard this term, but I see numerous misuses of it in the popular press, which typically interprets it to mean that people prefer gains over losses. Kahneman did not receive his Nobel Prize for such an obvious observation! Another misunderstanding is to conflate loss aversion with risk aversion, which means that people don't like to take risks—true but different, as explained in chapter 8 of this book. So let's make sure we understand what loss aversion is.

Traditional economists would say that the value of $100 remains the same whether you gain $100 or you lose $100. That seems to be a perfectly rational thing to think, as they are exactly the same amounts of money. So, if you find a

$100 bill in a dryer while you're washing your clothes, and it makes you happy by, say, 37 units on some hypothetical positive-negative mood scale, then losing a $100 bill because it fell out of your pocket should make you less happy by 37 units. But Kahneman and Tversky claimed that $100 feels different to us when we gain it versus when we lose it. Here is a demo to illustrate this.

Suppose I invite you to play a simple game. I toss a coin, and if it lands on heads, I give you $100, but if it lands on tails, you give me $100. Would you play this game? Almost everybody says no.

Now let's make this game a bit more attractive. If it's tails, you give me $100, but this time if it's heads, I give you $130. To be a bit fancy, we can calculate what is known as the expected value for this gamble. The chance that you will lose $100 is 50 percent and the chance that you will win $130 is also 50 percent, so the expected value is 0.5 × (-$100) + 0.5 × $130, which is $15. That is, if you play this gamble over and over again, you will sometimes win and sometimes lose, and the average payoff you can expect in the end is $15. That's greater than nothing, so a rational person who thinks like a mathematician, statistician, or economist should choose to play this game (assuming they want to make money). But again, only a few people are willing to play a game with these rules. I certainly wouldn't. I could definitely use $130, but if I have to give up $100 in cash just because of the way the coin behaves, that would be deeply tragic, much more tragic than getting a parking ticket because the meter expired five minutes ago. So I would let go of that opportunity, as most people did, even if it is worth $15.

It's not until the win/loss ratio is at least 2.5:1 (that is, you win $250 for heads and lose $100 for tails) that a majority of people will play. This is loss aversion. Loss looms much larger than gains. People weigh the impact of negativity far more heavily than the impact of positivity.

To translate this into real-life investment decisions, let's say Alex is presented with an opportunity to invest $10,000. Let's also say there are only two possible outcomes. There is a 50 percent chance that this would grow to $30,000 in one year, and she will end up gaining $20,000 within a year. But there is also a 50 percent chance that Alex will never see that $10,000 again! That sounds catastrophic. As a result, Alex turns down this opportunity when the expected value is significantly positive; namely, $0.5 \times \$20,000 + 0.5 \times (-\$10,000) = \$5,000$ of gain. Calculating expected values like this may make it seem easy to avoid the negative effects of loss aversion when you need to make a decision. As we will see next, however, loss aversion can manifest in less tangible ways.

Suppose you decide at long last to get rid of your old car and buy a new one. You spend a month researching, pick a make and model, and visit a dealer. You and your husband have agreed that the exterior color you want is "celestial silver metallic" and the seats are "ash" leather. You think you're all set. But then the salesperson starts asking you about all sorts of options, like auto-dimming mirrors, blind-spot alerts, "evasive steering assist," and so on. He says the base model is $25,000, but you can add feature X at $1,500, and feature Y at $500, and so on, and so on. Every time he presents a feature, he explains how it would make your life so much better and safer—that is, what you would gain.

At a different dealer, a savvier salesperson proceeds in the opposite direction. She starts out with a fully loaded model at $30,000. Then, she says if you give up feature X, which could save your life, the price would be $28,500, and if you also lose feature Y, which could make your parallel parking so much easier, it would be $28,000. This salesperson frames your choices in terms of the features you would lose. And that pushes your loss aversion button.

Does it work? In a study conducted in the 1990s, participants were told to imagine one of the two situations I just described. Those who started with a base model of $12,000 (prices were a lot lower back then) and were asked to add features (framing the choice in terms of what can be gained, or the gain-frame) spent $13,651.43 on average. In contrast, participants who started with a fully loaded model of $15,000 and were asked which features they were willing to lose, ended up spending $14,470.63 on average, about $800 more than those who received the gain-frame. If we convert this to the current car price of, say, $25,000, it would be like spending $1,700 more just because the price were presented in loss-frame.

Most of the studies I've cited took place in labs, and the

decisions or judgments were about imaginary situations, so skeptical economists defending the rational model of human behavior could dismiss them as not replicable in everyday life, where the stakes are real. Interestingly, some of the researchers who made this point carried out what they called "field experiments" in the real-life settings of urban K–8 schools in Chicago Heights, a city thirty miles south of Chicago. These involved not just hypothetical scenarios with hypothetical money, but real money, namely teachers' salaries.

You might have heard of teacher incentive programs, in which teachers receive merit pay if their students do well on standardized exams. A typical method is to give teachers a bonus at the end of the year, after the students have taken the tests. In the study carried out in Chicago Heights, some teachers were randomly selected to be in the "gain" condition, in which this traditional method was adopted; they received a year-end bonus, depending upon their students' improvement. Based on the reward ratio the researchers preestablished, its expected value was $4,000.

Another group of randomly selected teachers received $4,000 at the beginning of the year. This group was in the "loss" condition, because if their students' performance at the end of the year was below average, they would have to return the difference between $4,000 and the bonus they were actually entitled to.

The researchers ensured that the teachers involved would receive the identical net payments for a given level of student performance, regardless of whether they were in the gain or the loss condition. What the researchers were seeking to measure was whether the difference in the timing of the bonus would affect how incentivized the teachers would feel, and whether that would cause a difference in the students' performance. That is, would the average student grade improve over the prior year with a teacher who was trying to earn a bonus or with a teacher who didn't want to lose a bonus—or in both cases, or neither?

In the gain condition, the incentive program had essentially no effect. This wasn't the first time such an incentive program failed; the same result was observed in a study conducted in New York City. An end-of-the-year bonus—at least for the amount used in that study—was simply not sufficient to motivate the teachers.

In contrast, the scores of the students whose teachers were in the loss condition improved by as much as ten percentiles. It seems that not wanting to give up money was a very powerful motivator for the teachers. But, of course, the only difference was the timing of the payment!

Although the results are impressive, we will need to wait and see whether this study brings about any changes in public policy, or if it even should—it's possible the teachers in the loss condition were motivated to simply "teach the tests" or otherwise game the system. But on a smaller scale, we can think about ways to use the same technique to motivate others or even ourselves.

One summer I offered my son money to paint our deck.

It was a large sum for a high school student who had just graduated, and he readily agreed to do it. The summer passed and the only task he had accomplished was to mail-order brushes, rollers, roller pans, and a pressure washer. While painting the deck myself on a hot late summer day after it became clear to me that he wasn't going to do it before he left for college, I wondered why I didn't pay him cash up front but tell him he would have to pay me back the money if the deck went unpainted.

Maybe I didn't set it up that way because it feels rude and even cruel to take money away from someone after you've given it to them. I can't imagine paying my hairstylist a tip up front and demanding the money back if I wasn't happy with the cut. And think of the stress that the teachers in the "loss" condition at the Chicago Heights public schools must have felt whenever their students didn't do so well on a test. They must have felt as if they were under a constant threat of losing money; schoolteachers don't make much, so it's a fair bet that the extra money they got had been used to pay down bills or buy things they really needed. But that is precisely what the irony is: we are talking about the same $4,000, but it feels more threatening if we might lose the gain than if we never had it in the first place.

Endowment Effect

Loss aversion can also help explain why buyers and sellers seldom agree on the value of an item when negotiating the price. Let's say Annie is looking for a used stationary bike and has found a three-year-old one that was originally purchased for $300. Annie thinks it's worth $100; even though it looks brand-new, it *is* a three-year-old model, after all. The bike owner, Jenny, thinks it's worth $200 because she hardly used it. This is a very familiar scenario for any transaction of used items: the owner thinks it is more valuable than the buyer does. In behavioral economics, the phenomenon is called the endowment effect.

The pricing mismatch can occur simply because the seller wants to make as much money as possible and the buyer wants to pay as little as possible. The owner may also have a sentimental attachment to the item. But above and beyond those factors, the endowment effect occurs because of the mere fact of ownership and the instinct we all have to avoid losing something that is ours, no matter how briefly we might have had it—more specifically, because of loss aversion. The endowment effect arises instantaneously, even before any sentimental attachment can be formed, as demonstrated in the following clever study.

In this experiment, undergraduates were given a choice between a mug with the logo of the college they were attending and a Swiss chocolate bar. About half the students chose the mug and the other half the chocolate bar. This was just a baseline condition to establish what percentage of these undergraduates might prefer one over the other.

Then a different group of students from the same school was presented with the same choice, the mug versus the chocolate bar. But this time, the procedure was modified in a seemingly trivial way. The students were first offered a mug and they were told they could keep it. Then they were asked whether they'd like to trade the mug for the Swiss chocolate bar. This essentially is the same as asking whether they want the mug or the chocolate. So, about half of them should trade given the results of the baseline condition. But only 11 percent chose to exchange the mug for the Swiss chocolate.

To make sure there's nothing special about starting out with a mug, students in the third group first received a chocolate bar and were asked whether they would like to trade it for a mug. The same thing happened. Even though about half should trade the chocolate for the mug, only 10 percent were willing to do so; 90 percent preferred to stick with the chocolate.

What is particularly remarkable here is that the students who were given the mug or the chocolate bar had no time to get emotionally attached to it. Nor were they trying to profit from any of this; it was clear that the mugs and chocolate bars had only a modest resale value, if any. Nevertheless, once they owned the mug, exchanging it meant they would be losing it. Ditto for the chocolate bar. People simply hate to lose what they own, even when they owned it for only a brief moment.

Oddly, one study showed that the pain of losses is literally physical. Participants took 1,000 milligrams of either acetaminophen or a placebo, and then filled out an unrelated

survey for thirty minutes—long enough for the acetaminophen to start working on those who took it. Then half the
participants were given a mug and were told that it was
theirs to keep (endowment condition), and the other half
were presented with a mug that was described as the prop-
erty of the laboratory (no endowment condition). Finally,
all participants, regardless of the conditions or whether they
had taken acetaminophen or not, were asked to list their
price for the mug should they sell it. Those who took a placebo exhibited the endowment effect; their selling price was
significantly higher in the endowment condition than in
the no endowment condition. But those who took the acetaminophen did not; their selling price was statistically no
different whether they were endowed with the mug or not.
It would be amusing if Tylenol added this warning to its
list of side effects: "Acetaminophen may cause you to ignore
losses and sell your possessions at a lower price than usual."
Or if the FDA allows it, they may start advertising, "Can't
ditch a partner who won't commit? We're here for you," or
"Want to sell your house fast? Take Tylenol."

WHY THE NEGATIVITY BIAS

As with many cognitive biases, the negative bias is with us
because it was and still is useful. Some scientists have argued that this bias may have been especially necessary early
in human history because our ancestors lived so close to the

margins of survival, where losing something meant dying, so they had to prioritize the prevention of potential losses. When you can't afford to lose anything, additional gains are a sort of luxury. To give a modern-day analogy, it's like driving a car on a highway when the gas tank indicator arrow is pointing at "E," the bright red "no gas" warning light has been on for the last fifteen minutes, and you know that the next exit is ten miles away. If you are in that situation, you don't mind turning off the AC, even if it's scorching hot outside, because you can't afford to waste even a drop of gas.

We now live in more affluent environments in which most of us do not have to treat every loss as a direct threat to our existence. Yet the negativity bias still plays a very useful role, because it draws our attention to things that need to be fixed. We don't need to constantly attend to what is going well. For instance, we are typically not conscious of our breathing or walking; we take these activities for granted as long as they work. And that is a good thing, because we shouldn't waste energy thinking too much about things we can do without difficulty or pain. When, however, our breathing becomes labored or our walking difficult, it's time to take an action. The threat of losing our ability to breathe or walk is a powerful motivator. Similarly, when we are about to lose possession of something, our attention should be directed to it. A grade of C or D isn't just a grade; it's a signal to a student that they need to pay more attention to their schoolwork. An innate form of the negativity bias can be found in parents, who are hardwired to respond to negative signals from their babies, such as crying, or an unusual color or smell in whatever their babies discharge. What keeps parents up at night is not the babies' cute smiles or soft skin, but their crying and vomiting. It is a biologically built-in negativity bias for our offspring.

COSTS OF THE NEGATIVITY BIAS AND WHAT WE CAN DO

Even though negativity biases served a purpose for humans and may still in some situations, they can be harmful when they become extreme. For instance, if parents' sensitivity to their children's problems extends beyond their early childhood, it's a recipe for teenage drama. *Are you done with your homework? What happened to your face? Why don't you exercise more?* And because these biases may be hardwired, merely being aware of them can't always help us avoid falling prey to their more harmful incarnations. Still, we aren't entirely powerless. There are ways we can counteract the negativity bias. Here are two possible strategies, one in the context of making wrong choices because of loss aversion, and the other in the context of the endowment effect.

The most obvious cost of the negativity bias is that it can lead us to make the wrong choices. We might miss buying a book that could have changed our life, simply because we let a few negative reviews cancel out the dozens of raves. Or we might pass on an investment opportunity that is a great bet based on the expected value because we're too worried about the possibility of losing some money.

One method that can be effective in cases like these is to take advantage of another cognitive bias known as the framing effect. Our preferences and choices are based on how the options are framed rather than the options per se. I already described some examples of the framing effect earlier in this chapter. One was that we may take flights that are on time 88 percent of the time but avoid flights that are delayed 12 percent of the time; another featured the salesperson who presented the all-in price on the car and subtracted features in contrast to the less successful salesperson who started with the bare-bones price and attempted to add to it.

The framing effect is so powerful that it can literally be a matter of life and death. When patients with lung cancer were told they had a 90 percent chance of surviving if they underwent surgery, more than 80 percent of them opted for the operation. But when patients were told they had a 10 percent chance of dying after surgery, only half of them chose the intervention. Clearly, patients should be presented with both framings so their decisions are not swayed by either the negativity bias or the positivity bias.

Taking this framing effect one step further, we can also try to reframe the questions we ask ourselves. Here's a study that illustrates this. Participants read about a hypothetical custody battle between Parent A and Parent B, who were in the midst of a messy divorce. Participants learned the kinds of details about the parents that are relevant for custody decisions, as shown in the table below. Parent A is average on all these dimensions, not great but not bad either. In contrast, Parent B is more mixed with some positive features, such as "above-average income," and also some negative features, such as "lots of work-related travel."

PARENT A	PARENT B
average income	above-average income
reasonable rapport with the child	very close relationship with the child
relatively stable social life	extremely active social life
average working hours	lots of work-related travel
average health	minor health problems

One group of participants were asked which parent they would *deny* custody to. The majority of them chose Parent B. That kind of makes sense. After all, Parent B has lots of work-related travel and some health problems, albeit minor ones. An extremely active social life can't be good for the kid either, they probably reasoned.

The other group of participants were posed with the same question, but framed in the opposite way; to which parent would they *award* custody? The majority of the participants in this group chose Parent B. That also makes sense given that Parent B has a very close relationship with the child and above-average income. But it means that across the two groups, Parent B was judged to be both better and worse than Parent A.

Searching for reasons to deny custody, people focus on negative features and neglect positive ones. Searching for reasons to award custody, they focus on positive features and neglect negative ones. (If this reminds you of a study on the confirmation bias covered in chapter 2 on the effect of "Am I Happy?" versus "Am I Unhappy?," it should, because exactly the same mechanism is at work.) Thus, when you feel like you are being overly bothered by negative features, you may find a more neutral balance if you frame the question in a positive manner—not just which option you would reject, but which you would choose.

Now, let's consider how we might avoid the endowment effect. The endowment effect can lead us to make wrong choices when mere ownership causes us to place a greater value on something than it's actually worth. One example is when we fall for marketing tactics that exploit it. Free trial memberships are one of the most common. We know it's free for thirty days, and we enter the end-date in the calendar to remind us to cancel the membership, so it feels benign. But the endowment effect makes the membership feel a lot more attractive once we own it; suddenly we feel we can't do without this thing we never even really wanted.

My family got a Disney+ membership just to watch the movie version of the Broadway show *Hamilton*. Although it wasn't a free trial, the monthly membership was only $6.99, and the show was totally worth it. And it'd be a piece of cake to cancel it. Or so I thought. Rationalizations for keeping the subscription started to seep in after we watched *Ham-*

ilton three times. Who knows, we might be in the mood to watch the *Star Wars* series again or maybe *Frozen* . . . and it's cheaper than the price of a scone and a venti latte at Starbucks.

Another example of a sales tactic that relies on the endowment effect is the "free return" policy. Because we know we can get our money back if we don't like the item, we are more likely to take a risk and order it. Once it arrives and especially after trying it on, all of sudden having to repack it and drop it off at the post office feels like a daunting task. Even if we are not in love with it, we say, "OK, I kind of like it, I think I can find an occasion for it." So much for that no-risk free return.

Which brings us to our closets. The endowment effect and loss aversion are certainly the top reasons why some of our closets are so cluttered. Parting with clothes we haven't worn for more than three years can be as painful as parting with an old friend. We may still remember how much we paid for those treasures. Or even worse, who gave them to us. We never run out of excuses to keep them. My husband has six pairs of ragged pants and three pairs of old shoes all saved for gardening, which he barely finds time to do on more than a few weekends a year. I myself held on to a big-shouldered Armani jacket that I snatched from an 85 percent off clearance rack in the 1990s, and a couple of pencil skirts from the B.C. (Before Children) era.

Then I read *The Life-Changing Magic of Tidying Up*, the no. 1 *New York Times* bestseller by Marie Kondo. She is a professional organizer, not a psychologist, but no one understands loss aversion better than she does. To overcome this fear, the first thing she commands us to do is to take everything out, all the clothes on hangers, everything in the drawers, and all the shoes on the racks, and dump them on the floor. Because we just dumped them all, we don't own them anymore. No more endowment effect and nothing to lose. As a result, our decisions are reframed as deciding which ones to select, changing a loss-frame into a gain-frame. We can now evaluate each item on its own merit rather than out of the fear of losing it. When I Marie Kondo'd my closet, pretending that I was purchasing each item from the giant pile, the decisions were straightforward. I would never buy a skirt that is one size smaller. Nor a big-shouldered jacket, even if the style might come back ten years from now.

And the free trial subscription and free returns? Having already watched *Hamilton* three times, I asked myself whether I would start subscribing to Disney+ now if I had to start a new subscription. And I pretended that the dress I ordered online is something that needs to be ordered, except I now know that what looked like rose pink on my computer screen is actually fuchsia. The membership was canceled, and the dress has been returned.

6

BIASED INTERPRETATION

Why We Fail to See Things As They Are

IN 1999, I WAS PREGNANT WITH MY daughter and meticulously preparing for her arrival. The due date was in early June, and by May I was all set with the essentials: a car seat, two strollers, eight receiving blankets, fifteen bibs, ten boxes of diapers, ten onesies. I was beginning to tackle items I deemed as only somewhat less pressing, including books like *Goodnight Moon* and *The Very Hungry Caterpillar* (I believe in early education) and a night-light, when I ran into a study in *Nature* that made me reconsider buying a night-light.

Babies who slept with a light on in their rooms were five times more likely to become nearsighted than those who slept in the dark, the study said. It received quite a bit of attention in the mass media. As CNN summarized it, "Even low levels of light can penetrate the eyelids during sleep, keeping the eyes working when they should be at rest. Taking precautions during infancy, when eyes are developing at a rapid pace, may ward off vision trouble later in life." Of course, I crossed night-lights off my ever-expanding list of newborn prep.

One year later, another paper was published in *Nature* debunking this earlier study. It turned out that the correlation between night-lights and myopia was due to the parents' eyesight. Parents who have myopia were more likely to use night-lights, and because of hereditary factors, the children of nearsighted parents were more likely to grow up to be nearsighted themselves. CNN duly corrected its earlier report: LEAVE IT ON: STUDY SAYS NIGHT LIGHTING WON'T HARM CHILDREN'S EYESIGHT. This is a great example of how correlation does not imply causality, but that is not where I am going with this. Bear with me a little longer.

In 2001, a year after the night-light study was debunked, I became pregnant with my son. Given what I knew by then, would I—a person with severe myopia—use a night-light in his room? The answer was *absolutely not*. It felt wiser to risk bumping my knee into a corner of a dresser or stubbing my toe against a trash can than to take even the smallest risk of damaging my children's precious eyes. (Of course, both of my children wear glasses, despite all the bruises I suffered.)

As a cognitive psychologist, I became interested in this resistance. I even came up with a name for it: causal imprinting. Here is how it goes. At first, let's say in Stage 1, a person observes a correlation between A and B, as shown in the figure: when A is present, B tends to be present, but when A is not present, B also tends not to be present.

Based on this observation, one could infer in Stage 2 that A causes B, like night-lights cause myopia. The critical phase is Stage 3. At this point, one learns that there was also a third factor, C, and that whenever A and B co-occurred, C was also present, and whenever C was absent, A and B did not co-occur. Based on this observation, the most valid causal inference would be that C causes A and B, and A does not cause B. The observation of the correlation between A and B in Stage 1 is spurious because C was as yet unknown. Nevertheless, once someone has been imprinted with the belief that A causes B, they still interpret the common cause pattern of data in Stage 3 as A causing B, even after they learn about C, and even when there is no evidence that A causes B in the absence of C.

Stage 1: Observe	Stage 2: Infer	Stage 3: Observe	Stage 4: Infer	Correct Answer:
A ···· B	A → B	A ···· B ˙C˙	A → B ↖C↗	A B ↖C↗
and		**and**		
not-A ···· not-B		not-A ···· not-B ˙not-C˙		

In a series of experiments I carried out with Eric Taylor, who was working with me as a postdoctoral fellow, we found that participants who started out at Stage 3 (that is, by observing all A, B, and C factors at once) easily got the correct causal relationship, which was that C causes A and B, and A does not cause B. So, it's not as if people have inherent difficulties with learning a common-cause structure.

But just like me with the night-lights and myopia, if participants start out from Stage 1 and develop the initial belief that A causes B, that belief is imprinted and does not get revised, even after seeing the full pattern of data that clearly indicates that the causal association between A and B is false. Once one believes that A causes B, nothing in the new set of data presented in Stage 3 directly contradicts that belief: A and B still seem to be co-occurring, so one interprets that correlation as evidence that A causes B and does not revise their faulty belief.

This is another example of confirmation bias, our tendency to stick with our preexisting beliefs. In chapter 2, I discussed the type of confirmation bias that occurs when we don't *seek* information that might contradict what we already think is correct. This time, the confirmation bias happens because we *interpret* new data to fit with what we think is the truth.

THE PERVASIVENESS OF BIASED INTERPRETATIONS

Here is another story involving my child and biased interpretation. When my second child was four years old, we had a debate while I was driving. He asked me why a yellow traffic light is called a yellow light. I didn't understand his question, but he was only four, so I said, "It's called a yellow light because it's yellow." To which he replied, "It's not yellow, it's orange." I patiently corrected him, while wondering whether my husband had neglected to tell me that he is color-blind, and so might have passed this trait on to our son. My son insisted, "Mom, just look at it." To prove that he was wrong, I stopped at the next yellow light and stared at it. And there it was: an orange light. OK, it's not the color of ripe Florida oranges, but it is undeniably closer to the color of an orange than a lemon. Just look at one for yourself. Later I learned that yellow traffic lights are made orangish on purpose, to ensure maximum visibility (in officialdom and in the UK, they are referred to more accurately as amber lights). Fine, but then why did I grow up thinking they were yellow? I felt like I'd been deceived all my life. My parents called them yellow lights and I have always called them yellow lights. When I was little, I dutifully drew traffic lights with red, green, and lemony yellow crayons. The scariest part is that until my son corrected me, I actually *saw* them as lemony yellow.

Having a biased interpretation of reality because of what we already believe is extremely common. Though this example is not inherently dangerous—it doesn't really matter whether you call a traffic light yellow, orange, or amber as long as you obey it—you would assume that people will revise their initial beliefs in the light of new data when the consequences of not doing so are detrimental. Yet there are many examples of biased interpretations that persist in the face of contrary evidence, even when they can harm both oneself and others in significant ways.

For example, we all know at least one person who always attributes the blame for their problems to someone else. When they are late for a meeting, they blame it on traffic, despite the fact that there's traffic on that road at that time every day. When they hurt someone's feelings, they apologize by saying, "I'm sorry you felt that way." Believing that they are always right and that someone else is always wrong may protect their fragile egos, but it deprives them of opportunities to learn and grow, and to develop strong, healthy relationships.

And then there are those who blame everything on themselves. They feel compelled to doubt any compliment they receive ("He must say that to everyone"), to downplay their achievements ("I was lucky"), and to amplify even the most constructive negative feedback as damning ("I am hopeless"). Perhaps they suffer from impostor syndrome. They're never good enough, and no new evidence to the contrary can ever break through their preexisting negative notions about themselves.

People who suffer from depression are particularly prone to biased interpretations that are harmful to themselves. Suppose Ella texts her friend Les: "Any plans for Friday night?" Four minutes later, the message status changes from "delivered" to "read," but Les has not replied. Now it's been two hours. There could of course be many, many reasons Les

hasn't replied. He could have just walked into an excruciatingly boring meeting that made him forget about the text,
he could have dropped his phone into a giant bowl of noodle
soup right after reading it, or a bird might have pooped on
his head and he's been washing his hair with antibacterial
shampoo ever since. Although the situation is completely
ambiguous, Ella, who has been feeling doubts about her
worth, concludes Les doesn't want to be friends with her
anymore.

People also hurt others when they perpetuate inaccurate
impressions about them that are based on unwarranted ste-
reotypes. There are countless studies showing this, but one
of my favorites examined the gender pay gap, a troubling
and controversial societal problem. Women are paid less
than men, but some argue that is not unfair because it reflects true differences in aptitude. The study I'm about to
cover examined what happens when two candidates for a
research job are identical in every respect except gender.

The participants in this experiment were science professors at large American universities with highly promi-
nent, well-respected science departments. They were asked
to rate a candidate for a position as a student laboratory
manager. The application indicated where the candidates
had received their bachelor's degrees, their grade point av-
erages, their GRE scores (which is like the SAT except for
graduate school), their previous research experience, future
plans, and other information typically asked of people applying for jobs. All the professors who participated in this

study were presented with the same application except that on half of them the applicant's name was Jennifer, and on the other half, it was John.

Despite the fact that Jennifer and John's credentials were exactly the same, the study's participants, all of them science professors who had been trained to interpret data without any biases, rated John to be significantly more competent, hirable, and deserving of faculty mentoring than Jennifer. When asked to estimate how much salary they would offer the applicant, the average for John was more than $3,500 (or 10 percent) higher than that for Jennifer. These scientists interpreted the same application in different ways solely because of the gender of the applicant. It is all the more disheartening that this was true not only for the male professors making the judgments but also for the female professors.

Countless similar studies demonstrate bias based on every kind of -ism one can think of, not only sexism, but racism, ethnicism, classism, heterosexism, ableism, and ageism. Let's consider one that examined a set of atrocious problems that have received a great deal of attention recently: police violence and racism. Participants—mostly white male and female—were asked to play a video game in which an individual pops up unexpectedly in real-life scenes (like in front of a mall or at a parking lot) holding either a gun (a silver snub-nosed revolver or a black 9 mm pistol) or some other object (like a silver-colored aluminum can, a black cell phone, or a black wallet). The researchers took pains to make these objects clearly identifiable on the screen and not ambiguous. Participants were instructed to "shoot" the person if he is holding a gun, and to press a not-shoot button if he is not holding a gun. They had to do this under time pressure, simulating the kinds of situations that police officers confront when called into potential crime scenes. As readers may have already predicted, the target person was sometimes a white man and sometimes a Black man.

And you have probably guessed the chilling results. Participants were significantly more likely to shoot a Black man without a gun than a white man without a gun. That is, an aluminum can is more likely to be mistaken for a silver revolver if a Black man is holding it. In addition, participants were significantly more likely to misjudge a white man with a gun than a Black man with a gun. That is, a black pistol is more likely to be interpreted as a black cell phone or a wallet if a white man is holding it.

In one of the follow-up experiments, the researchers examined how fast the participants pressed the "don't shoot" button when the target was unarmed. This time, they made sure to recruit not only white participants, as in the earlier experiments, but also Black ones. Both white and Black participants pressed the "don't shoot" button faster when the unarmed target was a white man than they did when the unarmed target was Black.

SMART PEOPLE
CAN BE MORE BIASED

Are some people less susceptible to bias? How about those who are typically considered smart? We might like to think that people who are more intelligent can discern what is right or wrong and apply only relevant knowledge to help them interpret data or judge what they see. Conversely, when we hear that some people react to certain events in a manner completely opposite to how we believe they should, it's tempting to think of them as less smart than us. For instance, suppose there is a person who strongly believes that COVID-19 is no more fatal than a regular flu. We might think that only foolish people could believe such a ludicrous theory and treat the deaths of millions of people around the world as "regular" deaths, believing that they were all about to die anyway. But there are plenty of people who have demonstrated intelligence in other parts of their life who parrot this demonstrably false idea.

In fact, smarter people can be even more prone to biased interpretations, because they know more ways to explain away the facts that contradict their beliefs. A seminal study, published in 1979, is probably the most frequently cited in writings about confirmation bias, especially the kind that can lead to political polarization. But the fact that it required elaborate and intelligent efforts on the part of the participants to maintain their bias has not been much commented on, so here are the details.

Undergraduate students were recruited to participate in the study based on their views on capital punishment. Some were supporters of the death penalty, as they believed it deters crimes. Others were opponents of the death penalty. Upon entering the lab, the participants were asked to read the findings from ten studies that examined whether the death penalty increased or decreased crime rates. Half of these (hypothetical) studies showed the deterrent effect, as in this one example:

Kroner and Phillips (1977) compared murder rates for the year before and the year after adoption of capital punishment in 14 states. In 11 of the 14 states, murder rates were lower after adoption of the death penalty. This research supports the deterrent effect of the death penalty.

The other half reported that capital punishment did not deter crime rates:

Palmer and Crandall (1977) compared murder rates in 10 pairs of neighboring states with different capital punishment laws. In 8 of the 10 pairs, murder rates were higher in the state with capital punishment. This research opposes the deterrent effect of the death penalty.

Every time the participants read a study, they were asked to rate changes in their attitudes toward the death penalty.

By this point, readers may expect that I'm about to report the same old confirmation bias: that supporters of the death penalty said they still had positive views of capital punishment, while opponents of the death penalty still had negative views, regardless of what studies they read.

Interestingly, that's not quite the case. After reading the results of a study showing the deterrent effect, both the supporters and the opponents became more positive about capital punishment. Similarly, both groups became more negative after reading the opposite results. That is, people were affected by new information even if they contradicted their original views. Their initial attitudes moderated how much change was observed—for example, after receiving the deterrence information, proponents become even more positive about capital punishment than opponents did—but it did not prevent people from making some adjustments.

Critically, the study had a second phase. Having gone through only the brief summaries of results, now the participants were asked to read more detailed descriptions of the studies, which fully fleshed out their methodological details, such as how the states for these studies were selected (since U.S. states have different laws), or the length of time the study covered. Participants also learned exactly what the results looked like. These details made a huge difference, because they provided these smart participants with excuses to dismiss the evidence when the results contradicted their original beliefs.

Here are some examples of what the participants said:

*The study was taken only 1 year before and 1 year after
capital punishment was reinstated. To be a more effective
study they should have taken data from at least 10 years
before and as many years as possible after.*

*There were too many flaws in the picking of the states and too
many variables involved in the experiment as a whole to
change my opinion.*

Using such elaborate critiques, they convinced themselves that the studies whose results contradicted their initial
beliefs and attitudes were flawed. Not only that, the contradictory results made them even more convinced about their
initial position. Supporters of capital punishment became
even more positive about capital punishment after having
read the details about the studies that undermine the deterrent effects of capital punishment. Similarly, opponents of
capital punishment became even more negative about it after having read details about the studies that supported the
deterrent effects. That is, evidence that contradicted their
original beliefs resulted in even more polarization.

Coming up with excuses to dismiss evidence requires
a good amount of analytic thinking skills and background
knowledge, like how to collect and analyze data, or why
the law of large numbers, covered in chapter 5, is important. When the participants could not apply such sophisticated skills because the study descriptions were so
brief, biased assimilation did not occur. But once they had

enough information, they could use those skills to find fault with the studies that contradicted their original position, to such a point that findings that were at odds with their beliefs ended up strengthening them.

This study, however, didn't directly investigate the individual differences in the reasoning skills of the participants. Another study more directly examined whether individuals at different levels of quantitative reasoning skills differ in biased interpretations. The researchers first measured participants' numeracy, their ability to reason using numerical concepts. The questions they used to measure numeracy varied in difficulty, but they all required a fairly high level of quantitative reasoning to answer correctly—some just a little more complicated than, say, calculating tips, or computing the price of a pair of shoes during a 30 percent discount sale and some much harder to figure out. Questions like these:

Imagine we are throwing a five-sided die 50 times. On average, out of these 50 throws how many times would this five-sided die show an odd number? (Correct answer: 30)

In a forest, 20 percent of mushrooms are red, 50 percent brown, and 30 percent white. A red mushroom is poisonous with a probability of 20 percent. A mushroom that is not red is poisonous with a probability of 5 percent. What is the probability that a poisonous mushroom in the forest is red? (Correct answer: 50 percent)

Then, participants were presented with some "data" showing a relationship between a new skin cream and a rash. The table below shows what the participants saw. In 223 out of a total of 298 cases (or about 75 percent of the time) when the skin cream was used, the rash got better, and in the remaining 75 cases, the rash got worse. Based on such data, many people would jump to the conclusion that the new skin cream makes the skin condition better.

But remember how I used monster spray and bloodletting to illustrate confirmation bias in chapter 2? Just like we must check what happens when we don't use monster spray, we also need to look at the cases in which the new skin cream was *not* used. The data summarized in the table below show that in 107 out of 128 cases (or about 84 percent of the time) when the new skin cream was not used, the rash got better. In other words, according to this data, those with a rash would have been better off not using the skin cream.

	RASH GOT BETTER	RASH GOT WORSE
Patients who did use the new skin cream (Total = 298)	223	75
Patients who did not use the new skin cream (Total = 128)	107	21

Assessing these results correctly is a fairly challenging task, so it makes sense that the higher the participants' scores were in the numeracy assessment, the more likely they were to get the right answers. And in fact that's what happened. I should also add that there was no difference between Democrats' and Republicans' abilities to make the right assessment. This may seem like an odd thing to mention here, but it was critical to establish, because in another condition of the study, participants received numbers that were identical to those used in the skin cream and rash data, but these were now presented in a politically charged context.

This data was about the relationship between gun control (specifically, prohibiting concealed handguns in public) and crime rates. Two versions of this data were presented: one showed that gun control increased crime, supporting the view held by a majority of Republicans, and the other showed that gun control decreased crime, supporting the view more common among Democrats.

Whether the participants were Democrats or Republicans, those who scored low in numeracy still had trouble getting the correct answers, as with the skin cream and the bloodletting examples; they were at a chance level determining whether gun control increased or decreased crime. At least, their interpretations of the data were not biased. Regardless of whether the data showed gun control as increasing or decreasing crime, Democrats and Republicans with lower numeracy scores were more likely to be incorrect than correct, and there was no difference

REPUBLICAN-CONSISTENT DATA

	DECREASED CRIME	INCREASED CRIME
Cities that did ban carrying concealed handguns in public (Total = 298)	223	75
Cities that did _not_ ban carrying concealed handguns in public (Total = 128)	107	21

Note: According to these fictitious data, gun control increases crime because 25 percent of cities with gun control showed an increase in crime whereas 16 percent of cities without gun control showed an increase in crime.

DEMOCRAT-CONSISTENT DATA

	DECREASED CRIME	INCREASED CRIME
Cities that did ban carrying concealed handguns in public (Total = 298)	75	223
Cities that did _not_ ban carrying concealed handguns in public (Total = 128)	21	107

Note: According to these fictitious data, gun control decreases crime because 25 percent of cities with gun control showed a decrease in crime whereas 16 percent of cities without gun control showed a decrease in crime.

between Democrats and Republicans, just as with the skin cream version.

Among those with higher numeracy, however, there *was* a bias. Republicans with higher numeracy were more likely to get it right when the correct answer was that gun control increased crime. Democrats with higher numeracy were more likely to get it right when the correct answer was that gun control decreased crime. That is, people with stronger quantitative reasoning abilities used them only when the data supported their existing views.

I am not trying to say those lacking high levels of quantitative or analytic reasoning skills don't make biased interpretations. Of course they do. It is highly unlikely that only "smart" people make, for example, snap race-based judgments about whether someone is holding a gun or a cell phone. The point here is that so-called smart skills do not free people from irrational biases. Sometimes they can exacerbate the biases.

WHY WE INTERPRET
FACTS IN A BIASED WAY

Interpreting facts and data to fit with one's own biased beliefs can easily become a menace to individuals and society. Before I discuss what, if anything, we can do to counter this tendency, it's worth considering why we have it and why we so often fail to recognize or counter it.

Undeniably, motivational factors play a crucial role. The

motivation can be the need to save face, to show that one is correct (even when we're not). Sometimes we also have a desire to protect the credos of the family, clans, or political parties to which these beliefs—and we ourselves—belong. Explaining biased interpretations in terms of motivated reasoning is valid in some cases. There are, however, many situations in which our interpretations are biased when there are no motivating factors. Consider the traffic light again. I had no vested interest in believing the middle light to be yellow. I have strong opinions about many issues, but the color of traffic lights isn't one of them. Despite that, I've been wrongly seeing those lights as yellow since I was a child, simply because I believed them to be that color. Or consider the female science professors who offered a lower salary to Jennifer than to John; it's hard to imagine that they actively wanted to block women from the field of science. Few would believe that those Black participants, who were quicker to decide not to shoot a white unarmed target than a Black unarmed target, wanted a more racist society. Even when we are not motivated to believe something, our existing beliefs can color what we see or experience because that is how our cognition works. Recognizing these biases as part of our cognitive mechanisms can help us grasp how deeply rooted they are.

The cognitive mechanisms behind interpretative biases are no different than those we use at every moment of our lives. Humans possess a vast amount of knowledge that we constantly, unconsciously, and automatically use whenever we process external stimuli. In cognitive science, this is called top-down processing.

For instance, consider how we process audio input, like someone saying something. Those who grew up in the United States must have repeatedly heard the Pledge of Allegiance: "I pledge allegiance to the Flag of the United States of America, and to the Republic for which it stands, one nation under God, indivisible, with liberty and justice for all." It's not unusual to hear children reciting it as "under God, invisible" or "to the Republic for witches stand" because it actually sounds like that. When we consider the purely phonetic properties of those words, such errors are understandable. It is only when one thinks about the true meaning of the pledge that we realize that it can't be "witches" or "invisible."

Think about voicemail transcription features. I've been quite impressed with the way my iPhone dictates phone numbers for me, and the recent messages it has transcribed have been remarkably accurate. Despite this impressive improvement in artificial intelligence, I got a voicemail transcription last week that read: "Hi this message is for _____ on my name is Mary I'm calling from yell at your nose and throat please give our office a call back at [number redacted] and it's option number three again I'm calling from Yale your nose and throat." I can let go of the fact that the transcription failed even to attempt my name (unless there is a systematic cultural bias built into the AI system), but what is "yell at your nose" and "Yale your nose"? I played the actual voicemail from Mary, who enunciated "Yale Ear

Nose and Throat" with as much clarity as one could possibly imagine. But the fact is that speech is highly ambiguous, and it's thanks to our top-down knowledge and the wealth of references that we draw on unconsciously that we are able to disambiguate words and meanings as well as we can. My top-down processing was so strong and automatic that I absolutely could not hear "yell at your nose" no matter how many times I played the recording.

But can people see an identical object in two opposite ways just because of what they believe at the moment, even if they have no vested interest in what they believe? My former graduate student Jessecae Marsh and I examined that question in an experiment. At the beginning, each participant was shown a slide that had a picture of a kind of bacteria that was present in a soil sample on the left side of the screen (it looks like a bar) and next to it a picture of the sample (as seen in the figure panels). The sample was clearly labeled to indicate whether nitrogen was present in it or not. The participants were told that they would be shown a number of such slides, and would then have to figure out whether a certain kind of bacteria causes the presence of nitrogen in the soil.

Each participant then saw sixty screen pages, each showing on the right a different soil sample. At first, they were shown two kinds of bacteria, as seen in the top two figure panels: in some, the bacteria stretched from the top of the image to the bottom, and in others it appeared as a small bar with lots of space above it and below.

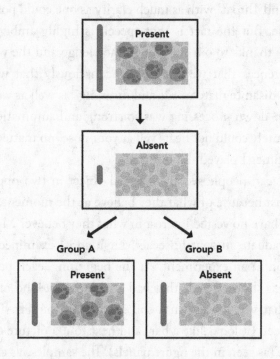

As illustrated in the top two boxes of the figure, participants first saw several soil samples where extremely long bacteria were paired with the presence of nitrogen, and extremely short bacteria were paired with absence of nitrogen. After seeing several such pairs, participants were likely to develop the belief that long bacteria cause the presence of nitrogen in the soil. That's all pretty straightforward.

Then we added a twist. Around the time that participants started thinking that long bacteria cause the presence of nitrogen, half the participants, we will call them Group A, were shown soil samples where medium-length bacteria were paired with a picture that indicated the presence of nitrogen. The length of this medium bacteria was carefully constructed to be right in between the obviously long and obviously short bacteria. That is, if you were categorizing the length of the bacteria as long or short, these medium ones were truly ambiguous, neither long nor short. Medium.

By the end of the experiment, Group A had been shown a total of sixty slides, a mixture of long bacteria paired with soil that had nitrogen, short bacteria next to soil that showed *no* nitrogen, and medium bacteria paired with soil that had nitrogen. Then, the participants received a surprise question: How many slides showed long bacteria paired with samples that had nitrogen? Out of the sixty samples, only twenty showed the obviously long bacteria, the one that went from the top of the slide to the bottom, and those were always paired with soil that had nitrogen. But on average, the participants reported having seen about twenty-eight cases. Because they had developed the hypothesis that long bacteria tend to cause nitrogen production, whenever they saw a slide with nitrogen in the soil, they interpreted the bacteria as being long, even if they were sometimes seeing the completely ambiguous middle-sized bacteria.

The other half of the participants in the same experiment, Group B, saw a very similar sequence, starting with long bacteria (nitrogen present) and short bacteria (nitrogen absent) images, but this time the medium bacteria they were shown in the later slides were paired with the *absence* of nitrogen. When asked to estimate how many slides they had seen in which short bacteria were paired with the absence of nitrogen, they estimated it to be twenty-nine on average. The correct answer was twenty.

In other words, both groups of participants saw identical medium-length bacteria in the second part of the experiment, but Group A saw them as "long" and Group B saw them as "short" because both groups had come to believe that long bacteria cause the presence of nitrogen. Given that initial belief, they interpreted ambiguous bacteria accompanying a slide of nitrogen-rich soil as long and ambiguous bacteria alongside nitrogen-less soil as short. I'm absolutely certain that none of our participants cared about this belief. It's not like they were going to make money if they saw more long than short bacteria or vice versa. Furthermore, they didn't have to count those medium ones; they could have just ignored them because they were ambiguous. But they spontaneously classified them as "long" or "short" because it fit with their top-down view.

Not only did they categorize the bacteria this way, they also started seeing them this way. At the very end of the experiment, we presented the participants with pictures of all three bacteria and asked them which one the medium bacteria looked more similar to, the long or the short bacteria. Group A said that the medium bacteria looked more like the long one and Group B said it looked more like the short one.

Top-down processing takes place spontaneously and automatically, whether we are motivated to use it or not. We need it to make sense of the world, as it places information that comes to us through our senses into a coherent framework, allowing us to predict and control our environment. We would be completely lost and our lives would be chaos without top-down processing.

Consider a very basic visual perception, like what I am passively seeing now as I type this sentence, which is my dog moving out of his bed. The physical features of everything in my visual field—the form, the color, the contours, the lines, the shapes—are constantly changing. But I see a single object (my dog) moving out of another single object (the bed) and stepping onto the floor, rather than either the dog bed or the floor transforming into different shapes and colors. Now imagine that instead of me the perceiver is a robot that is perfectly capable of processing physical signals using a super high-tech camera. But also imagine that this robot doesn't have the concepts of a dog or a bed, or the basic principles of perception. It doesn't know that parts that move together may well belong to a single object. Moreover, it doesn't understand more abstract notions like animacy: it sees all inanimate objects as capable of becoming animate and thinks that computer graphics exist in the natural world. Given all this, the robot wouldn't understand the scene in the same way as I would. Without top-down processing, we would be like that robot, unable to distinguish our dogs from their beds and constantly expecting appliances and furniture to spring to life.

WHAT ARE WE TO DO?

The problem here is that top-down processing is also responsible for biased interpretations, which in turn cause confirmation bias and prejudice. The outcomes of such biases are often appalling, but the process itself relies on capabilities that we use all the time to make sense of the world. In other words, we cannot easily stop the process that gets us in trouble; we need it. Understanding that biased interpretations are inevitable is a good first step when we are figuring out what we can do to counter their perils.

Thinking biases are that much harder to overcome when we believe that we don't commit them, and that they only plague dense people who aren't like us at all. Once we realize that biased interpretations are a part of top-down processing, we can admit that we all are capable of and prone to committing biases in interpretation, even when we are trying to be open to thoughts not dictated by a certain doctrine and haven't been brainwashed by a wacky cult group. Keeping this in mind, the next time a four-year-old tells you that a yellow traffic light is orange, you may be more open-minded and give it a fresh look.

Unfortunately, solving life's problems is not always as easy as looking more closely at a traffic light: there is no simple fix when we hold incorrect beliefs about ourselves, believing, for example, that we are losers or our future is hopeless when that is not the case at all. Let's continue with this example. Everyone feels doubt about themselves at times, and some people have a particularly hard time shaking it off, so much so that doubt becomes a part of their self-concept. When that happens, they interpret everything that happens to them in light of that flawed belief, which further reinforces it. As a result, it becomes virtually impossible for them to free themselves from those doubts on their own.

In clinical psychology, there is a technique known as cognitive behavioral therapy that is specifically designed to de-bias deeply entrenched negative thinking styles. It might sound strange to some that we need to learn better ways of thinking (and even have to pay for it if one's insurance doesn't cover it), but we do. Here's one way to understand this: when we go to a buffet, we don't dump any random foods into our mouths like Pac-Man swallowing everything in front of him; we deliberately choose which dishes to eat and which to ignore. In the same way, there are always a lot of thoughts going through our mind and we need to select which ones to attend to and which ones to let go. If someone has formed the bad habit of indulging negative thoughts, that person will need help to break the habit, in the same way that we need a yoga instructor or a personal trainer at a gym to teach us exercise techniques and cheer us on so we use them consistently. Cognitive behavioral therapy has been shown to be highly effective, but like getting into shape with the help of a personal trainer or a yoga instructor, it does not work like a magic wand in a single session; it takes weeks and weeks of therapy sessions, and one has to also repeatedly exercise the skills in everyday life—another example of how challenging it is to counteract biased interpretations.

Let's switch gears. What are we to do when we are inconvenienced or burdened by someone else's biased interpretations? Again, understanding that such biases are partly cognitive may help us to be more tolerant with those who see things differently. That is, it's not always the case that these people wish to harm us; they might just see the situation in their own way. We don't have to get defensive every time. Sometimes, it's easier and better to focus on solving the problems caused by different perspectives than trying to change those perspectives themselves.

For example, let's say Mr. Green is obsessed with maintaining his lawn, while his neighbor Mr. Brown believes that manicured lawns are environmental hazards that require dangerous chemicals and waste water. What Mr. Green sees in Mr. Brown's garden are nightmarishly ugly, noxious, repulsive, invasive weeds, but what Mr. Brown sees is a beautiful and hardy assortment of indigenous wildflowers. When a similar conflict occurred in *The Great Gatsby*, Gatsby sent his own gardeners to take care of the neighbor's grass, but even if Mr. Green could afford to do that, that solution won't work given Mr. Brown's philosophical principles. Rather than arguing about whether manicured lawns are sustainable or not, Mr. Green might as well just plant some hedges to block his view of Mr. Brown's garden and redirect his obsession to keeping them well trimmed.

Yet, as we saw earlier in this chapter, the harms caused by biased interpretations go way beyond the level of eyesores in the neighborhood. Prejudices against certain groups can easily become matters of life or death. What are we to do when others hold views that are morally repugnant to us? We all know how challenging it is to change someone's worldview. Many of us have learned not to bring up politics over Thanksgiving dinner if we ever want to see certain family members again.

This is why we sometimes need policies and regulations at the systemic level. For instance, it is enormously challenging to convince someone to get a COVID vaccine if they believe that the vaccines are harmful. My friend's friend's friend has a Ph.D. in biology and an elaborate and totally false theory about how mRNA COVID vaccines permanently damage our genes. Even so, her daughter ended up getting vaccinated because her college required her to do so before she could return to campus. This is an example of how a change at the systemic level can protect public health, even when people's views are widely divergent. Similarly, the Equal Employment Opportunity Act of 1972, which addresses discrimination based on race, religion, color, sex, or national origin, is another systemic-level approach. Needless to say, we should continue educating people to counteract their prejudices; we need to de-bias people as much as we possibly can. But biased interpretations based on our beliefs about our health, intrinsic values, and safety are entrenched and often immutable once they are formed. Furthermore, many of those prejudices have systemic origins, caused by our history, culture, economics, and politics. System-level changes also come with their own challenges. For one, there is a recursive problem that these decisions are to be made by people who are also prone to biased interpretations.

Nevertheless, sometimes the only way to counteract one system is with another—one that is explicitly, equitably, and intentionally designed to protect the greater good.

7

THE DANGERS OF PERSPECTIVE-TAKING

Why Others Don't Always Get What's Obvious to Us

MY HUSBAND AND I ONCE ATTENDED a dinner party with two other couples. Our hosts are famous in our circle for creating and leading clever party games. On that evening, they introduced us to a wine-tasting game. Each couple was presented with four glasses, labeled A, B, C, and D, that had been filled with different varieties of red wine. One person in each couple was told to taste them and write down descriptions of their tastes on four index cards. The descriptions were to be just that, descriptions, and there was no indication on the card whether they applied to wine A, B, C, or D. Then the other partner would also taste the wines and try to match them up to their partners' descriptions.

One of the couples were avid wine connoisseurs. They own a large wine cellar and travel to wineries around the world. The husband tasted the wines and described them using the jargon that is familiar to wine experts: medium-bodied, oaked, austere, buttery, herbaceous. It was truly intimidating when his wife read his descriptions out loud. But after all that, she was only able to match one description correctly. It was a hard game.

The second couple were both English professors and the husband composed a little poem for each wine. He compared one to the valley they looked down on from the cabin they stayed in to celebrate one of their anniversaries, and another to the joy they shared when they overcame an ordeal. It was jaw-dropping that he could write such splendid poems on the spot, and his wife read them out loud in a beautiful voice and intonation. We were all going *ooh* and *aah*. But they got zero matches.

My husband and I had been married for about fifteen years at the time, and we are both psychology professors. People often ask us whether we can read other people's minds, and the answer we give them is no—if our line of work has taught us anything, it's how overconfident most people are when it comes to knowing what's in their own minds, never mind other people's. But my husband knows me really well, and one thing that he knows about me for sure is that I know nothing about wine. I have normally functioning taste buds, but I am just as happy with a cheap boxed white blend as an expensive vintage. Worse still, I don't even like red wine.

My husband took less than a minute to fill out his cards. I smiled when I read them, and matched them all correctly.

What he had written was: "most sweet," "second," "third," and "least sweet."

HOW BAD ARE WE AT COMMUNICATING?

We communicate with other people all the time. We talk or write about our new ideas or feelings, and we also listen to and read what they communicate to us. Despite doing this all our lives, we don't realize how difficult it is. The other couples at that dinner party were quite bitter after the wine-matching game, as it was inconceivable to them that their spouses could have not understood their perfect descriptions. The oenophiles complained that they underperformed *only* because the red wines had not been given enough time to breathe. Unfortunately, miscommunications are more common than we think, even among people that we know. I will first show two studies that powerfully demonstrate just how bad we are at communication.

We can begin with written communications, like emails or texts. We use texts with our friends and family to update what's going on, ask questions, or joke around. While doing so, many of us use sarcasm, like "so sad to have to miss that reunion" or "my boss has done it again." When we text sarcastic sentences, we assume the recipient knows we are being sarcastic. When we receive texts, we also assume that we're fairly good at detecting sarcasm. But is that really the case?

In one study, participants were tested on their sarcasm recognition abilities using sentences written by their own friends. The group was divided into pairs. One person was instructed to email their partner a series of one-sentence messages, some of them sarcastic and some serious. The senders were highly confident that their partners would know whether they were sarcastic or not; they were friends after all, who knew their dry sense of humor or lack thereof. The messages' recipients were also fairly confident about their judgments. Yet when the scores were tallied, their accuracy was merely at a chance level, that is, 50–50, no different than flipping a coin. It is scary to think that half of the sarcastic jokes that we've tweeted, texted, or emailed might have been taken as serious, and that half of our serious statements might have been mistaken for sarcasm.

If it makes you feel any better, you don't have to freak out about all the sarcastic things you've ever said in your entire life, because this result occurred only when the messages were delivered in writing. When similarly sarcastic or serious sentences were delivered through a voice message, people understood them as they were intended. That's because, at least in English, there is a fairly recognizable sarcastic intonation that speakers use, with slightly elongated syllables and a higher pitch. Most people recognize what that tone indicates.

That said, you can still freak out a little. Another study found that failures can happen in many cases even when we are using our voices and trying to match our tone to our intent. This study used ambiguous sentences that frequently crop up in everyday conversations, such as, "Do you like my new outfit?" When your partner or friend asks that question, it could be because she is worried that the outfit is not flattering, because she thinks the outfit is perfect and is fishing for a compliment, or because she's annoyed that you didn't even notice what she is wearing. In fact, there are many utterances that we use that are quite ambiguous when you think about them, such as "Please leave me alone." This can mean "I'm busy," or alternatively, "I'm mad at you." A simple question like "How's the salad?" could mean "Isn't the salad terrible?" or "Why aren't you saying anything nice about the salad I made?" Or, you could be literally asking how good or bad the salad is. And unlike with sarcasm, there really isn't an agreed-upon intonation for each of those different meanings.

In the study, one person in each pair of participants was provided with several sentences like those above and instructed to convey a certain meaning to the listener by saying them out loud. The listeners had to guess for each sentence which one of four possible interpretations the speaker intended. The listener was either a stranger the speaker had just met in the lab or a close acquaintance, like a friend or spouse. When they were spouses, the couples had been married for 14.4 years on average.

As in the sarcasm study, the speakers were confident the listeners had understood what they intended to convey. Not surprisingly, they were even more sure when the listeners were their friends or spouses. To the contrary, there was absolutely no difference between acquaintances or strangers in figuring out the intended message. On average, the listeners correctly guessed the intended meaning of fewer than half of the sentences. That is, even after fourteen years of marriage, your spouse may misunderstand your tone of voice, and hence the meaning of your potentially ambiguous sentences, as much as half of the time.

CURSE OF KNOWLEDGE

Obviously, nobody wants to misunderstand their friends and family. Nobody wants to be misunderstood either. So, why does this happen? Whenever we perceive something, we interpret it in the light of what we already know (as we talked about in chapter 6). Because we do that automatically and unconsciously, we may believe that everybody else, including even a person who doesn't know what we know, would see the situation similarly to the way we do.

Studies showing this egocentric bias have been carried out with young children. The classic task goes like this:

Sally has a marble. She puts the marble in her basket.
Sally goes out for a walk.
Anne takes the marble out of the basket and puts it into
    the box placed next to the basket.

Now Sally comes back. She wants to play with her
marble.
Where will Sally look for her marble?

The correct answer, of course, is in the basket, not the box. But most children under the age of four say that she would look for her marble in the box, because they know that's where it is. They have difficulty reasoning that others may have false beliefs, beliefs that are different from the reality that they know. If readers have heard the term "theory of mind," this is what it refers to: reasoning about what's in other people's minds.

Because the errors shown with children are so obvious, you might think adults wouldn't commit them, but a later study found that even college students have similar difficulties. Participants learned about a girl named Vicki. She was practicing violin in a room in which there were four containers of different colors. When Vicki finished practicing, she placed her violin in the blue container and left the room. While Vicki was outside, Denise came in and moved the violin to a different container. At this point, half of the participants are told that the container Denise placed the violin in was the red one (we will call these participants the knowledge group); the other half are not told which container Denise chose (the ignorance group). Then, participants in both the knowledge and the ignorance group are told that Denise rearranged the containers so that the red one is where the blue one used to be. Finally, they are asked to estimate for each of the four containers what the likelihood is that Vicki, upon coming back to the room, would look for her violin in it. The correct answer is obviously that she will look in the blue container. Nonetheless, participants in the knowledge group—those who knew that the violin was actually in the red container—could not completely ignore that information; they rated the likelihood that Vicki would search for the violin in the red container higher than those in the ignorance group did. That is the curse of knowledge: once you know something, you have trouble fully taking the perspective of someone who doesn't know it, even if you are an adult.

Those who have played the board game Pictionary must have experienced how it feels to suffer from the curse of knowledge. In this game, one person draws a card that has a phrase or a word written on it and sketches a picture representing it. Then the other members of the team have to guess what's written on the card based on the drawing. Let's say the sketch includes the face of a person with long hair. Apparently, this person is a she, because the artist has given her breasts. Next to her there are four smaller people, also with long hair and breasts. What the heck is this a picture of? When the time runs out and nobody has guessed correctly, the person who drew it exhibits the curse of knowledge, yelling at the teammates, "How could you have missed it when it's sooo obvious? It's *Little Women*, four daughters with their mom."

In the next round, a person claiming to be a better artist pulls their card and draws the face of a lion. A teammate shouts out "lion," but that is not the correct answer. Others on the team ask the artist to add something to the drawing. But the artist keeps pointing to the elaborately drawn lion's face, as if she's saying that they clearly do not need anything more than that. Another person guesses, "Mane!" Nope. The artist points again, stabbing the picture with her pen so hard that the paper gets punctured. That's how frustrating the curse of knowledge can be. And still nobody guesses. (By the way, the answer is the Chronicles of Narnia.)

Of course, Pictionary is meant to be challenging because it's a game, and furthermore, not everyone is good at drawing. Here's a famous study that hardly requires any skills. In fact, readers can try this at home or wherever they are if they can find someone who has two minutes to kill. Participants are asked to select a well-known song that any partner randomly paired for the experiment could recognize. Let's say Mary is one of the participants and she chooses "Mary Had a Little Lamb." Mary taps the song without singing it. And the partner has to guess what it is.

Readers can tap the song of their choice now. This feels like almost anybody should be able to guess it, right? In the actual study, the tappers estimated that about 50 percent of their listeners should be able to correctly guess the song they were tapping. But did 50 percent of listeners get it right? Well, it felt easy only because the tappers knew what song they were tapping. In the study, 120 songs were tapped and only three were correctly guessed. The tappers were experiencing the illusion that anybody could guess their song, simply because the answer was playing in their minds.

If your partner has another two minutes to spare, ask them to pick a song to tap for you, so that you can experience what it feels like to be on the other side. When we do this exercise in my class, one of the most common wrong answers is Queen's "We Will Rock You" because it actually starts with tapping (or stamping) and no melody. Even "Happy Birthday" can sound like hard rock.

The curse of knowledge makes us overconfident about the transparency of the messages we are conveying. For instance, the tapper might make a small rhythmic error, which can completely throw the listener off. But the tapper may not think it's a big deal since they tapped exactly the same phrase only seven seconds earlier. Again, the tapper assumes the listener can hear the music that's playing in their head, just as the artist in Pictionary, who is seeing a mental picture of the book jacket of the Chronicles of Narnia with Aslan's picture on it, can't imagine what else she could possibly add to her sketch to make it clearer.

The wine-tasting game that started this chapter also illustrates the curse of knowledge. The advantage that my husband and I had was our lack of confidence. My husband, who clearly knew how unsophisticated—OK, ignorant—I am about wine, had no choice but to use what we might call wine vocab for dummies, which turned out to be the best strategy.

In fact, smart people who know a lot are not necessarily good teachers or coaches, partly because of the curse of knowledge. I have heard complaints from college students about a course taught by a Nobel laureate—absolutely brilliant, but utterly incomprehensible. One of my former students took violin lessons from a maestro who had won several Grammy Awards. When I asked her whether he was a good teacher, she tactfully answered, "Violin comes naturally to him."

FORGETTING TO CONSIDER THE OTHER PERSPECTIVE

Often our communication failures occur simply because we neglect to consider the other person's perspective. I'm talking about truly absurd cases, where we already know what the other person knows, thinks, sees, or likes, not the kinds of situations I've been describing up till now, where one person couldn't have known what the other person was thinking. Furthermore, in the cases I'm about to discuss, our actions depend on what the other person has in their mind, so we have to take that into consideration. But even in such cases, we can forget to think about the other person's point of view.

An example of this phenomenon is the status-signal paradox. Participants in a study are told to consider the following scenario:

Imagine you've just moved to Denver, and you're going out to a social activity at a downtown bar. You really want to make some new, close friends. As you're getting ready, you're trying to decide on which of two watches that you own that you should wear. One is an expensive designer watch and the other is an inexpensive generic watch. Both match your outfit. To what extent would people be attracted to become friends with you if you wore the designer watch? How about a generic watch?

If you chose the designer watch, you were like the majority of the participants in the study. The same results were found when using a Saks Fifth Avenue versus a Walmart T-shirt, a BMW versus a VW Golf, or a Canada Goose versus a Columbia coat. Just like peacocks showing off their iridescent feathers, humans want to signal their high status to other people by displaying luxury items, like a handbag with PRADA printed on it, a Rolex with its iconic crown logo, or a bright-red Ferrari with falcon-wing doors.

What is paradoxical, however, is the result from another group of participants. They had been recruited from the same pool (so they probably had similar tastes and values as the first group) but randomly selected to be asked a different question, namely to which person they would be more attracted. Their answers were the opposite. They would rather be friends with someone wearing a generic watch than a Rolex, a Walmart T-shirt than a Saks Fifth Avenue one, or someone driving a VW Golf as opposed to a BMW.

When we are choosing what to wear to appear attractive to potential friends, we can get trapped in the egocentric perspective of wanting to signal our high status and make the wrong choice. How would we feel if a potential new friend showed up at a bar wearing a Tag Heuer watch or a black T-shirt with GUCCI printed on it in gold? Only if we pause for a moment and take the opposite perspective would we know which watch to select. Even when we are trying to impress someone—or maybe *especially* when we are trying to impress someone—we can't forget to consider their perspective.

This next study also shows how people forget to take other people's perspectives into account, even when they should be able to. It also suggests that the culture you grew up in may have something to do with it. Participants were undergraduates from the University of Chicago who were told that they'd be playing a communication game. Each was seated across from a "director," who was one of the experimenters. Set between them was a wooden frame, about twenty inches square and five inches deep, standing upright. Inside it were four rows of four evenly divided cells, as shown in the figure. Some of those sixteen cells contained small objects, like an apple, a mug, or a block. The participant's task was to move the objects around the cells, following the director's instructions. For instance, the director might say, "Move the bottle one slot to your left," and the participant would search for the bottle, pick it up, and move it the left. Both the participant and the director could see these actions, and the participant's only task was to follow the director's instructions.

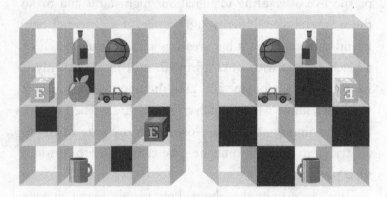

Note: The participant's perspective is shown left and the director's perspective is shown right.

After some warm-ups, the director then says, "Move the block one slot up." There was only one bottle and one apple, so there was no choice to be made. But this time, there are two blocks, as shown on the left side of the figure. Critically, one block—the one in the third row from the top—is hidden from the director's view (as shown on the right side of the figure). Participants could clearly see that the cell was occluded; they had even played the director role during one of the warm-up rounds, so they knew what it was like to be on the other side of the frame. Given these facts, the participant should have been able to immediately tell which block the director was referring to—the one in the second row from the top, as it is the only one that the director could see.

The experimenters measured how long it took for participants to complete their tasks in each round. Then they compared the reaction time when the competing block (that is, the one in the occluded cell) was present against the reaction time in the round when there was no competing block. Although the answer should have been obvious, it took participants 130 percent longer to move the correct block than it did in the rounds where a competing block was not present. Furthermore, almost two-thirds of participants unabashedly asked, "Which block?"—sometimes more than once. Some people were so oblivious to the director's perspective that they even moved the block in the occluded cell that the director could not see.

Interestingly, this confusion occurred only with native speakers of English. The researchers also tested University of Chicago students who were born and raised in mainland China and had been in the United States for less than ten months. When these Chinese participants were presented with the same task in Mandarin, their reaction times were identical, whether or not the competing block was present. In other words, they completely ignored the objects that were occluded from the director's perspective, and acted as if they were seeing exactly what the director was seeing. Only one Chinese participant asked, "Which block?," and I'm guessing they were quite embarrassed when they realized what they'd asked.

The reason for this cultural difference would make sense to those who understand the difference between collectivist and individualistic societies. Some cultures, including South Korea, Japan, India, and China, are known as collectivist. People in these cultures have developed a strong sense of belongingness, and are regularly reminded of their duties or responsibilities to the group. There's a constant attentiveness to social norms.

To give a simple example, consider ordering meals in a restaurant. In America, each person orders their own dish, and there is a tendency to avoid ordering the same dish that someone else has; if the first person to give their order to the server asks for what the next person was considering, the next person says, "Oh if you're getting that, then I'll order this instead." If they really want the same dish, they may feel compelled to apologize for not being more original. In Korea and China, the default is for entrees to be shared. Even if people are ordering individual dishes for a lighter meal like lunch, once an elderly person or whoever is higher in the hierarchy orders their dish, everyone else tends to order the same thing.

Loyalty and conformity to the group are valued in collectivist cultures to the point of sacrificing privacy and individual rights at times. During the pandemic, almost all South Koreans followed the government's orders to wear masks and close their businesses. A leader of a religious group who held indoor gatherings and caused an outbreak had to prostrate himself on the ground on national TV and beg for forgiveness. A QR code-based entry log system was mandated at stores, restaurants, nightclubs, karaoke rooms, and any other high-risk areas. When positive clusters of COVID were detected, visitors to those locations were notified that they had to take a COVID test. That level of social conformity would be unimaginable in an individualistic society like the United States.

To fit into these collectivist societies, members have to be constantly aware of what others are thinking and how others think about themselves. The socialization that is required to conform to these norms begins at a very early age. Perhaps as a result of this constant training to read others' minds, people from collectivist societies become so proficient at taking others' perspectives that it is almost reflexive.

WHAT WORKS

How can we better understand what others think, intend, believe, and feel? The fact that those who grew up in collectivist cultures are better at this means that these skills can be taught and learned. But we can't move to a collectivist society or send our kids to live in one for several years just to enhance their ability to grasp others' thoughts and feelings. And as some readers might suspect, hypersensitivity to what others might be thinking also has its downsides. For those in individualistic societies, the tacit pressure put on those in collectivist societies to order the same dish as others in a restaurant may sound bizarre if not outlandish. Having our whereabouts constantly known to others, even during a public health emergency, can sound like George Orwell's *1984*. We have also seen reports that suggest that being overly conscious of others' opinions can lead to serious mental health issues; at the very least, it can render you vulnerable to bullying, whether in the real world or on the internet. Clearly, we shouldn't let ourselves get too carried away when it comes to trying to figure out what others think. Nonetheless, we do need to make sure that we have the essential, rudimentary level of understanding of what others are thinking that is necessary to carry out normal social interactions.

Let's begin with some solutions for young kids. Remember the trouble that two- to three-year-old children have in understanding that other people can have false beliefs, beliefs that are different from the reality they know? A study found that children in this age cohort can be taught to understand false beliefs in less than a couple of weeks. Interestingly, the study was done in the context of helping them learn how to lie. Note that lying requires the basic understanding that even if *we* know the truth, other people may not. Because two- to three-year-old children do not understand this, they can't lie.

In the study, three-year-old children first learned a game in which the experimenter hid a piece of candy under one of two cups. If a child guessed which cup the candy was hidden under, they could have the candy. Then, the child was asked to hide the candy, and the experimenter had to guess which cup it was under. After the child finished hiding the candy, the experimenter asked them where the candy was. Each child knew that they could keep the candy if the experimenter guessed wrong, but almost always pointed to the correct cup, the one that had the sweet. Despite having just hidden the candy, these three-year-olds falsely believed that the experimenter already knew where it was. They couldn't imagine that anyone could believe something that they knew wasn't true, so they almost always told the truth.

After the researchers used this type of game to make sure that the kids in their study could not lie in this game, the children received training across six sessions, spaced out over eleven days. The training consisted of several tasks. For example, the experimenter showed the children a pencil box and asked them to guess what was inside. Children would say pencils. Then, the experimenter opened the box and showed them that it actually contained something else, say, ribbons. Next, the experimenter asked the children whether they originally thought the box had ribbons, and whether someone else who hadn't seen inside the box would have thought it contained ribbons. If the children didn't give the correct answers (that is, no and no), which they mostly don't at that age because they lack an understanding of false beliefs, they were corrected with some feedback and the task was repeated. In another part of the training, the children were told stories that were full of vocabulary involving mental states (*like, want, feel*) and were asked to create sentences using those mental state vocabularies. When the children in the experiment finished the training, and were tested on the "hide the candy under a cup" task again, they now tricked the experimenter on almost every single trial!

Obviously, it's not nice to teach children how to lie and cheat, but that's not what the researchers taught; the children simply learned about understanding other people's mental states. As the researchers point out, to some extent to know how to lie is an important social skill. We would seriously worry about the mental health or social skills of a friend who did not get the logic of a surprise birthday party and assumed that the birthday boy would know about it the moment his friends started organizing it, even if no one told him about it. A surprise birthday party involves some dishonesty, but it is possible precisely because we know that others can have beliefs that are different from ours.

Most of all, what the children gained through the training is what is called cognitive theory of mind, the insight that other people can understand the world differently than we do. But if we are to have empathy or compassion toward others, we also need to have emotional theory of mind: a comprehension of the fact that people can have different feelings, and knowledge of what they are likely to feel under which situations.

This distinction between cognitive and emotional theory of mind is crucial in understanding psychopaths. Lying and cheating require us to understand other people's minds, and when it comes to cognitive theory of mind, psychopaths are almost as good at it as nonpsychopaths. That is, they are skillful at reading what others think and predicting how they would reason, which is why they are able to manipulate them. What psychopaths are lacking, however, is *emotional* theory of mind. They are callous, cold-hearted, and pitiless because they are oblivious to other people's feelings.

Emotional theory of mind—understanding others' feelings and having compassion toward them—can also be improved by carefully thinking through other people's circumstances. Let me illustrate this more concretely with a study that required its participants to think about Syrian refugees. As of 2016, there were 5.5 million Syrian refugees, a quarter of all the refugees in the world. Participants were asked whether they'd be willing to write a letter to the president—Barack Obama when the study was conducted—and ask him to admit Syrian refugees to the United States. Only 23 percent of the Democrats in the study said yes. But some of the study participants received additional special instructions before they were asked to write a letter; they were told to put themselves in one of those refugee's shoes: "Imagine that you are a refugee fleeing persecution in a war-torn country. What would you take with you, limited only to what you can carry yourself, on your journey? Where would you flee to or would you stay in your home country? What do you feel would be the biggest challenge for you?" The percentage of Democrats in that group who were willing to write a letter to the president was 50 percent higher. Vicariously putting oneself in another person's situation can increase prosocial behaviors. (The effect was weaker with Republicans, probably because they tend to hold anti-immigration views, not because

perspective-taking does not promote empathy among con-
servatives.)

WHAT DOESN'T WORK

So far, I've explained that it is possible to improve our understanding of other people's minds at both cognitive and
emotional levels. Yet there is an important caveat. The kind
of understanding that I have talked about is extremely basic. In the case of the Syrian refugees, their situation is so
devastating and atrocious that almost any human could
empathize. And any normally developing child learns that
other people's thoughts can be different from their before
starting preschool. Can we go beyond this rudimentary level
and recognize what others think or feel simply by trying to
imagine ourselves in their situations?

It feels like the answer should be yes. Precisely because
we believe that it is possible, we often complain to those
who are impervious to our needs, "Why can't you try to see
this from my point of view?" When we are tortured by a
boss who expects too much from us, we wonder how they
could have forgotten what it was like when they were at
the entry level. It doesn't feel like it's too much to ask for a
little understanding! But our intuitions here are wrong, or at
least not supported by evidence. A team of three researchers demonstrated through twenty-four experiments that our

ability to understand what others are thinking or feeling cannot be improved merely by perspective-taking.

By the way, this is the largest number of experiments I have ever seen reported in a single article. (It was actually twenty-five experiments; I'll get to the last one at the end of the chapter.) The reason they had to report so many was because the claim they were making was so counterintuitive. Furthermore, if any given study found no effect for perspective-taking, it could be for methodological reasons rather than because there really is no effect: Perhaps the participants didn't try hard enough, or the task was too hard. Maybe it was only impossible to understand other people's minds in that particular case. A fancy way of saying this is that the studies are trying to prove a null effect, and in social science, null effects are notoriously hard to demonstrate. As an analogy, suppose your mother declares that her favorite socks must have been thrown away because she checked "everywhere"—dresser, nightstand drawers, under the bed, laundry basket. But your father might say that's not "everywhere" and ask her to check inside your brother's dresser, inside her coat pockets, on the dog's bed, and between the sheets on the bed. Even after she's searched through all those places, your mom still can't claim she searched "everywhere." It's now easy to see that it's much harder to prove that the socks are not in the house than it is to prove that the socks are in the house. The same goes with proving null effects from experiments.

My reading of their paper is that the authors tried nearly everything, using numerous tasks. They used the false-belief task, in which participants should be able to imagine themselves in the position of someone who has a false belief about reality, even though the participants know what the reality is. Another famous test they used is called the "Reading the Mind in the Eyes" test; it was initially developed to study children with autism. A subject is presented with a picture of a pair of eyes and asked to select the emotional label that best describes them. (Readers can easily find this test online and take it for free; a similar test is used to measure one's emotional intelligence.) They tested people's ability to detect fake smiles or lies. Other tasks involved more realistic interpersonal interactions. Participants were asked to guess their partners' preferences among activities like bowling or doing the dishes, or to make predictions about how their partner would react to movies like *Casino Royale* and *Legally Blonde*, or to jokes that some people might find offensive but others would think were funny (like "What is the difference between a battery and a woman? A battery has a positive side" or "Why are men like strawberries? Because they take a long time to mature and by the time they do, most are rotten"), and to various controversial opinions (like "Police should use whatever force is necessary to maintain law and order").

All of these twenty-four experiments had two groups of participants: a control group in which participants were free to use any strategies to make guesses, and a perspective-taking group that was strongly encouraged to take the perspective of the other person, like the person whose eyes are shown in the picture, or their partner's when it came to guessing their preferences, reactions, and opinions. Participants in this second group reported that they had become less egocentric, and they believed that their perspective-taking must have increased their accuracy. But across all of these various tasks, their accuracy did not increase.

Even my husband and I, two psychology professors, have fallen into this trap. Here are the details for the readers' pleasure. I'm the main chef of the house, and since my husband frequently eats out because of his job, I, who can cook a variety of dishes, choose to cook the kinds of foods that he likes whenever he eats at home. And since my kids don't always like my husband's dishes, I often have to cook two different kinds of pasta for a single meal (spaghetti with Bolognese sauce for my kids and linguini with broccoli rabe and Italian sausage for my husband), or marinate two different versions of chicken (boneless and spicy for my son and bone-in and nonspicy for my daughter and husband). Before I continue, I also need to underscore that my husband is the most considerate and unassuming person that I know of, and he shares our house chores equally. He also knows me very well—remember, he knew exactly how to communicate with me about four different kinds of red wine. When our second child left for college and we became empty-nesters, I confessed that it was awesome to think that I would only need to cook one version of dinner from now on. My husband said, "Yes, and it's a good thing that you and I enjoy all the same kinds of food." I laughed hysterically to learn that my husband, who has always been so perceptive about what I think and feel, thought that I cooked fried chicken and Italian sausage because I like them. To the contrary, I could have easily become a vegetarian, and if nobody is home, I have pistachio ice cream and blueberries for dinner. Then I realized that I had never told him that! To make matters worse, for more than twenty-five years I had wrongly thought my husband was aware of the compromises I had made.

Although this example, along with the twenty-four experiments, illustrates that we can't get the facts right—like what another person's favorite food is—merely by taking the other person's perspective or being considerate, it is difficult to give up on the possibility that we can somehow learn how to better guess what others are thinking. In fact, there are psychotherapy techniques that teach people how to reassess their situations in a more objective, as opposed to egocentric, way in order to modify their destructive thinking styles. You might have also heard of training programs developed to improve emotional intelligence, such as learning to better identify emotions from facial expressions. These could help.

We also know that actors and fiction writers are exceptionally good at taking other people's perspectives, and they must have been taught and practiced to develop such skills. Not everybody can attend creative writing or acting classes, but can we at least understand others better by watching lots of plays or reading novels, which are all about seeing the world from someone else's perspective?

A study published in *Science* tested whether we can better discern others' thoughts or feelings by reading literary fiction. Participants read a couple of short stories (such as "The Runner" by Don DeLillo, and "Blind Date" by Lydia Davis) and excerpts from recent bestsellers (such as *Gone Girl* by Gillian Flynn and *The Sins of the Mother* by Danielle Steel). Afterward, they were run through the false-belief tasks and took the "Reading the Mind in the Eyes" test. The experimenters did find significant improvements. Subsequently, the study received quite a bit of attention and was widely cited. When I read the article, I found it quite difficult to believe because the participants read the stories for only a brief time. If it's that easy, why haven't we achieved world peace?

And as it turned out, the study was not replicable. A more recent study published in *Nature* evaluated the replicability of social science experiments published in *Nature* and *Science* between 2010 and 2015, one of which was the one I just described. They found no evidence of improvement in those tasks after reading fiction.

Still, and as I explained earlier, null effects are difficult to establish. One very plausible possibility is that the effect of reading novels is real, but it requires lots and lots of reading over many years. People in collectivist societies are better at guessing what others are experiencing because of a lifetime of immersion in the culture. Similarly, psychotherapy techniques and emotional intelligence training all require their participants to carry out exercises consistently for a very long time before they see any improvements. The same goes with actors and fiction writers: their talents at taking a reader's or audience's perspective are likely the outcomes of long practice, as well as plenty of mentoring and feedback from others.

WHAT CERTAINLY WORKS

There actually is something very concrete that each of us can do to improve our ability to grasp others' minds and to convey our thoughts more clearly. And it is simple: Stop letting others guess what we think and just tell them. Also, when texting sarcastic jokes, add emoticons like ¯_(ツ)_/¯ or 🙂.

Yeah, articulating our thoughts sometimes feels awkward and dull. It definitely looks uncool to spell out that we're joking. Yet it would be much more prudent to remind ourselves of how clueless we felt when someone was tapping a song for us. I would ask, "Do you like my new shirt?" only if I am genuinely curious about my friends' thoughts about the shirt and only if I can still return it if the consensus is that it's a flop, rather than as a passive-aggressive statement of "you are not paying enough attention to me."

Likewise, stop trying to read people's minds and feelings. If you are a compassionate and accommodating person, it is particularly hard to resist the temptation to guess others' thoughts. But study after study has shown us how disastrous this can be. The only sure way to know what others know, believe, feel, or think is to ask them. "Just ask" was the twenty-fifth experiment of the paper that I described earlier. Participants were given a list of questions about their partners. One group was instructed to take their partners' perspectives; the other was given five minutes to ask them questions before they were tested. Compared to the group of participants who were merely told to take their partners' perspectives, those who were allowed to ask did much better. Such a demonstration may seem blindingly obvious: of course, we can do well on tests if we know the correct answers. But that is precisely the point. You can't get the facts right unless you gather facts.

In order to accurately understand what other people think, feel, believe, or know, one must find out the answer directly from them. If you don't even know how funny or offensive your friends think sexist jokes are, you cannot correctly guess their attitudes just by imagining yourself in their positions. Because we project our own knowledge and feelings onto others, we are overconfident and believe we know what they think. As a result, we don't bother or we forget to verify whether our assumptions are correct. Gathering facts is the only sure way we have to understand each other.

8

THE TROUBLE WITH DELAYED GRATIFICATION

How Our Present Self Misunderstands Our Future Self

I RECEIVED MY PH.D. IN PSYCHOLOGY when I was twenty-five years old, several years younger than most Ph.D. recipients in my field. This was not because I was some kind of genius, but because I had a deadline that I had to meet. I certainly hadn't planned to finish so early when I arrived in the United States from South Korea at age twenty-one to begin graduate school. I was still learning to understand phrases like "For here or to go?" at McDonald's, and I was puzzled why my office mate found it so funny when in response to her question, "What brought you here?," I answered, "An airplane." My plan to spend the typical five to six years completing my Ph.D. was abruptly derailed at the start of my fourth year in the program, when my advisor decided to move to a different school. He told me that if I could finish my dissertation by the end of the year, he would bring me to his new university as a postdoctoral fellow, which was my dream research job.

I had always done well at school, but whipping up a doctoral dissertation from scratch in just one year was a formidable challenge and I had to work like crazy to do it, putting off all fun and pleasure. No movies, no parties, not even beer; I worked sixteen hours a day every day and essentially lived on Cracklin' Oat Bran, milk, and coffee. Even after that year, I endured all sorts of challenges and disappointments. I tell you all this to show you that I am perfectly capable of hanging on for rewards that are significantly delayed.

But I am also the least patient person I know. I reply to student emails within microseconds of receiving them, and I need immediate answers when questions cross my mind. When I come up with an exciting research idea, I don't email my graduate student; I text them or go to their office. And when I have an impulse to get a haircut, I take the first available appointment. As much I as I regret all the times that someone else has mangled my hair, having to wait until my favorite stylist is available is sheer torture. I demand results, answers, and rewards immediately.

DELAY DISCOUNTING

I started out with two apparently contradictory stories, but they are not contradictory, as I'll explain later. Before that, I want to first show you how impatient many people are. Here is a typical test to measure how we discount delayed rewards.

Would you prefer $340 now or $340 in six months? This one is a no brainer. Everyone prefers $340 now.

Would you prefer $340 now or $350 in six months? Most people still prefer $340 now.

Would you prefer $340 now or $390 in six months? In a typical experiment like this, the majority of participants still prefer $340 now rather than waiting six months to receive $50 more. Preferring $340 now over $390 in six months may sound reasonable considering inflation, interest rates, or investment opportunities. That is, wouldn't it be wiser to take the money now and do something with it that could yield a higher return?

The answer is no. Suppose you get $340 now, and you put it in a bank or invest it in a stock. Assuming an ordinary economy, you are likely to end up with only a little more money than $340 after six months—$10 or $15 more at the most. To turn $340 into $390 in six months, the annual return would have to be about 30 percent. That is much higher than any interest rate on the market.

Another possible argument is that we should take the money now because, who knows, something may happen in the next six months. The person who offered it to you may change her mind or die. You could die. Or a nuclear war could render paper money useless, unless you burn it to stay warm. Or your extremely rich aunt could pass away before the six months are up and leave all her money to you, eclipsing the value of the extra $50 you were waiting for. These examples are all highly unlikely; the point is that only in such rare cases would $390 in six months be worth less than $340 now.

Let's do one more exercise to illustrate how we discount future rewards in an irrational manner. Given a choice between receiving $20 now or $30 a month from now, most people choose $20 now. But given a choice between receiving $20 in twelve months or $30 in thirteen months, yes, you guessed it: most people choose to take the extra $10 and wait one more month for it. When we compare these two choice situations, the inconsistency is obvious. We are talking about exactly the same differences: $10 and one month. No matter how much $20 or $30 is worth to a person, if the person chose $20 in the first-choice situation, they should choose $20 in the second. But the one-month difference in the present feels much greater than the one-month difference in the future.

Of course, there is a limit to this phenomenon. If the choice is getting $340 now versus $340,000 six months from now, everybody can wait. And that must have been how I saw getting my Ph.D. I deemed the degree and the subsequent research job as *way* more valuable than any of the immediate rewards of having a social life or eating normally. I am sure everybody has had similar experiences and made sacrifices in the present for a large reward in the future; I am certainly not saying that people as a rule are incapable of delaying gratification.

Nonetheless, we tend to discount the utility of future rewards more often than is justifiable. Numerous experiments in behavioral economics, like the choice situations I just described, have shown that we don't seem to delay gratification enough. Let's now consider some real-life examples that show how irrationally we discount future rewards. And I will call this discounting of delayed rewards "delay discounting," as they do in behavioral economics.

Consider climate change. When we recycle to reduce waste, plant trees to capture carbon, or spend extra money to buy an electric car, we are not immediately rewarded with cleaner air, lowered sea levels, or happy polar bears. It takes years and decades for those benefits to emerge; some may only be experienced by future generations. Even when we know that the future rewards from reducing our carbon footprint are invaluable, we may not be sufficiently motivated to turn down our air conditioners or spend a substantial amount of money on solar panels today. But not doing so is analogous to accepting $350 now rather than $350 billion several decades from now.

And for all but those few of us who genuinely enjoy spending hours on a treadmill every day and crave salads with ancient grains, almost everything that we are told to do to stay healthy—like sticking to our New Year's resolution to work out five times a week, or stopping after the first glass of wine—requires choosing the delayed gratification of prolonged life over immediate rewards. Every time we yield to temptation, we show how much more powerful the second is than the first.

The delay doesn't have to be far in the future for our best intentions to lose out to whatever treat is at hand. At the end of a long, bad day, you may be craving for your favorite comfort food, pizza. You have the local pizzeria's number and they guarantee delivery within half an hour. If you wait just thirty minutes, you will be rewarded with hot, yummy slices. Then, you see a bag of potato chips on the counter, which you know will ruin your appetite for the pizza that would be a much better reward for all the stress you endured that day. And you start nibbling on the chips anyway and end up feeling just plain annoyed with yourself.

Delay discounting applies not just to discounting future rewards but also future pains, which explains why we procrastinate. Many of us can live in complete denial of the existence of an unappealing task until just hours before the deadline or even past it. The pain of a dreary task in the future somehow feels much more manageable than exactly the same pain if it's happening now. So we put things off. In an attempt to prevent my students from procrastinating on their final papers until the night before they were due, I once asked them to list the pros and cons of starting an assignment at the last minute. They gave all the typical "correct" answers for what is wrong with procrastinating, like you never know what might happen at the last minute, and we tend to underestimate the time it will take us to complete a task. But I was more interested in how they would defend procrastination. Some of them claimed better performance:

Diamonds are made under pressure.

The stress and adrenaline caused by an impending deadline can result in increased motivation.

You can ponder longer, incubating your ideas about the assignment until that moment.

There were also arguments for efficiency:

Parkinson's law: work expands so as to fill the time available for its completion.

You can't get bogged down by details or perfectionism.

You cannot procrastinate it anymore.

One of my favorite claims rationalized procrastination by using material we had covered in my class: "You cannot commit the planning fallacy at that point."

WHY WE CAN'T WAIT AND HOW WE CAN LEARN TO

Those were all examples of how people irrationally discount the values of future outcomes. In order to avoid such situations, we need to consider why they happen. There is more than one reason. I'll present several, each accompanied by a way to counteract it.

Lack of Self-Control

Sometimes we cannot delay gratification because we lack impulse control. The smell of bacon when you're "hangry" can make you forget all about the benefits of a healthy diet. Pushing oneself to start a project now that is not due for another six months requires an enormous amount of self-control when there are episodes of your favorite show just waiting to be binge-watched.

One of the earliest studies on delayed gratification and impulse control, famously known today as the marshmallow test, was conducted with children in the 1970s. Three- to five-year-olds were presented with a single marshmallow and told that the experimenter was about to leave the room. They could eat the marshmallow right away, but if they waited until the experimenter returned, they would get an extra marshmallow. If they could not wait, they would not get a second marshmallow.

Readers may want to search for the marshmallow test on YouTube, because it will surely make your day; the kids are colossally cute as they try to resist the temptation to gobble up the immediate, lesser reward. Staring at the marshmallow, their eyes get crossed. Some of them sniff it, some touch it and lick their fingers, some poke at it as if they want to make sure it's real.

As anyone who has raised children or been close to them can easily predict, the time they can wait varies. Some hold out for fifteen to twenty minutes; others give in much sooner. But that's not why the marshmallow test became so famous. A stunning finding emerged more than a decade later. It turned out that the children's wait times predicted their verbal and quantitative SAT scores: the longer a child could wait for a second marshmallow when they were tiny children, the better they did on their SATs at the end of high school. (Some of you may have read in the popular media that a subsequent study debunked the marshmallow test, but that's not quite true. The correlation between the wait time and the SAT scores in that second study was still positive, albeit smaller, and that follow-up study was later convincingly critiqued for various methodological and conceptual reasons.)

If more good things come to those who wait, how can we help children resist the temptations of immediate gratification? This question was actually what motivated the original marshmallow study. The easiest way was to hide the white, fluffy, sugary marshmallow from the children's view while they were waiting. Additionally, if the children had a toy to play with or were told to think happy thoughts, their wait time also substantially increased, even when the marshmallow was kept in plain sight.

The same irrational impulse to take a lesser reward now and the same technique for thwarting it exists throughout nature. Distractions can help pigeons delay gratification. In case you want to know how researchers discovered this, here are some details. The experimenters first maintained pigeons at 80 percent of their free-feeding weight, which made them highly motivated to seek food. The pigeons learned that if they instantly peck a button on the front wall of their enclosure when it lights up, they will get their less-preferred "Kasha grain" right away, but if they wait for fifteen to twenty seconds before they peck, they will get their more-preferred "mixed grain" pellet. Pigeons are no more patient than humans; they overwhelmingly chose the immediate gratification of the plain Kasha grain rather than waiting for the better mixed grain. It's hard for pigeons to wait while doing nothing.

But the pigeons could wait if they were distracted. In another experiment, there was a second key on the opposite wall of their enclosure, which lit up at the beginning of the trial just like the first one. The pigeons learned that if they pecked this second key twenty times, which takes much more time and effort than pecking the first key just once to get the Kasha grain immediately, they would be rewarded with the mixed grain pellet. It turned out the pigeons were remarkably better at waiting the fifteen to twenty seconds for their preferred mixed grain if they could distract themselves by pecking the second key.

Resisting immediate temptation is tough. When a person enjoys one or two cocktails or glasses of wine with dinner every evening, it can require an enormous amount of willpower to break the habit. Still, if kids and pigeons can distract themselves from immediate temptation, perhaps we grownups can do it too. Drinking a delicious nonalcoholic beverage is easier than staring cross-eyed at your dinner partner's drink.

Havoc of Uncertainty

Our judgments involving future rewards or pains can become irrational because we have difficulty thinking through uncertainty. I'll explain by sharing one of my favorite studies. Although it isn't really about delayed gratification, it powerfully demonstrates how the feeling of uncertainty can mess up our judgments.

A group of students were asked to imagine that they took a tough exam and just found out that they passed. Then they were told to imagine they were offered a very attractive Hawaiian vacation package at an exceptionally low price that was only good for one day. They were then given three options; they could buy it, not buy it, or pay a $5 nonrefundable fee that would extend the special deal. The majority of participants opted to buy the vacation package now. That made sense; they'd passed the exam, so they had something to celebrate.

Another group of students were presented with the same choices except that they were told to imagine they failed the exam and had to take it again in a couple of months. A majority of the participants in this group also wanted to buy the vacation package now. That made sense too: they had two months to prepare for the exam, so why not go to Hawaii to recharge?

The results from the first two groups established that these students were generally inclined to take the vacation package, regardless of how they fared on the exam. Yet when a third group of students who hadn't been told how they'd done on the exam was presented with the same choices, a majority said they would pay the $5 so they could wait to make their decision until they'd found out their results. People were willing to pay extra so they could make their decision after the uncertainty was removed—even though they'd likely be making the same decision no matter what the outcome was.

Uncertainty about significant future outcomes can immobilize our decision-making. While waiting to see if you are going to be offered a job after an interview, or to find out if a business deal is going to go through, it's difficult to do anything, even things you typically enjoy. As Election Day drew near in 2020, I found it nearly impossible to work on anything, including a writing project I had committed to completing by the end of November. In the spirit of the Hawaii vacation package study, I thought through each of the possible outcomes. If Trump is elected, do I still have to do this writing? Yes. If Biden is elected, do I still have to do this writing? Yes. This allowed me to carry on with my writing even on Election Day. In fact, it was a refreshing distraction for me.

Although I was able to keep calm and carry on through uncertainty, if there had been a way to find out the election results sooner, I would have been willing to pay not just $5 but a handsome amount of money. Most of us want to reduce uncertainty as much as possible. This aversion toward uncertainty is normal but it can lead us to become unreasonable when our choices are between a certain and an uncertain outcome, as in the case of delayed gratification.

To explain, let's return to the preference for receiving $340 right now versus $390 six months from now. Money aside, this can be perceived as a choice between certainty and uncertainty, because the future is always uncertain. Who knows what might happen in six months? Most of our worries about not getting the $390 are irrational, as I mentioned before, because the likelihood of any of them actually coming to pass are extremely small. The problem, however, is that even when we know that the probabilities of, say, our dying in the next six months are tiny, they seem much larger when they are contrasted with something we know for certain. This is fittingly called the certainty effect.

There is a famous phenomenon in behavioral economics called the Allais paradox, which occurs because of the certainty bias. It's named after Maurice Allais, who received the 1988 Nobel Prize in economics. Allais was both a physicist and economist, so we need to talk about numbers, but they're money numbers so they're easy to understand.

Here's how it goes. Consider the first situation. You are offered two too-good-to-be-true gambles, and you need to choose only one:

Gamble A: 100 percent chance of winning $1
    million
Gamble B: 89 percent chance of winning $1 million,
    10 percent chance of winning $5 million, and 1
    percent chance of winning nothing

Which one would you choose? Take your time making a choice (but don't try to calculate the expected values that I introduced earlier in the book; follow your intuitive judgment).

I know I would definitely choose A over B. A million dollars is plenty; I'd happily take it and retire. If I chose Gamble B and won nothing, I would be kicking myself for the rest of my life. There seems to be no point in taking a risk with Gamble B, even with the 10 percent chance of winning $5 million. The majority of people choose Gamble A as well. Gamble B's 1 percent chance of winning nothing compared to Gamble A's 0 percent chance of winning nothing feels like a HUGE difference.

Now consider a second situation. Select one of the following less amazingly great but still pretty good gambles:

Gamble X: 11 percent chance of winning $1 million
and 89 percent chance of winning nothing
Gamble Y: 10 percent chance of winning $5 million
and 90 percent chance of winning nothing

Given these options, most people select Gamble Y. I would too. Even though I said I'd be happy with $1 million, if the difference in the probability of winning $5 million versus $1 million is only 1 percent, why not risk that small probability for $4 million more?

But wait a minute. If you choose A in the first situation and Y in the second situation, you are being inconsistent.

Let's go back to Gambles A and B. To select the better one, a rational person should cancel the component that is the same between the two options. Here are the two options presented in a slightly different way to make this cancellation easier:

Gamble A: 89 percent chance of winning $1 million
PLUS 11 percent chance of winning $1 million
Gamble B: 89 percent chance of winning $1 million,
10 percent chance of winning $5 million, and 1
percent chance of winning nothing

Both A and B have an 89 percent chance of winning $1 million, so we cancel it out. Then, what's left for these two gambles, which we will call A' and B' respectively, are:

Gamble A': 11 percent chance of winning $1 million
Gamble B': 10 percent chance of winning $5 million

Now, which one would you prefer, A' or B'? Probably B'. Note, however, the A' versus B' choice is identical to the X versus Y choice. I'm copy-pasting it for you here:

Gamble X: 11 percent chance of winning $1 million
     and 89 percent chance of winning nothing
Gamble Y: 10 percent chance of winning $5 million
     and 90 percent chance of winning nothing

As in X versus Y, most people would choose B'. But when it was A versus B, most people chose A, behaving inconsistently and irrationally. That's why this is called a "paradox."

The reason this happens is because the same 1 percent difference feels vastly different when we are comparing 0 percent to 1 percent and 10 percent to 11 percent. Mathematically, they are identical 1 percent differences, but psychologically, we treat them completely differently, because the first is the difference between something absolutely not happening versus some chance that it can happen—namely, the difference between certainty and uncertainty. In contrast, 10 percent and 11 percent both feel like small chances that aren't that different.

The Allais paradox is precise and beautiful (at least to me), but it does feel artificial. Behavioral economists tend to discuss choice situations using gambling examples, but it makes the phenomenon a bit less relatable. Why would anyone ever offer you a gamble that gives you a 100 percent chance of winning $1 million? That's not even a gamble, and nothing like it would ever happen in real life. Here are some real-life examples from the pandemic that we've all experienced.

According to the CDC, as of June 2021, the Pfizer-BioNTech vaccine was deemed 95 percent effective against serious illness of hospitalization due to COVID-19 and the Moderna vaccine was 94 percent effective. There have been plenty of complaints, concerns, arguments, and overreactions expressed about the COVID-19 vaccines, but I have not seen anyone complaining about the 1 percent difference in efficacy between the two vaccines. I received Moderna. While it annoyed me that I had to wait four weeks to get my second shot as opposed to three weeks if I'd gotten the Pfizer vaccine (remember how impatient I am?), the difference in efficacy did not bother me at all. Hypothetically speaking, if the government decided that the Moderna vaccine would be free but the Pfizer vaccine would cost $100 because of the difference in efficacy, very few people would pay for that 1 percent difference.

Yet if the Pfizer-BioNTech vaccine were 100 percent effective and the Moderna vaccine were 99 percent effective, it might be a different story. We would be talking about a 100 percent guarantee of not getting COVID-19 versus some chance of getting it. People might start paying more than $100 to get the Pfizer-BioNTech vaccine. That is the certainty effect.

Whenever we are faced with a choice that involves delayed gratification, there is the possibility that our preference for certainty (getting it now) over uncertainty (getting it in the future) will be a factor. It is not easy to overcome this hypersensitivity. I've been teaching the Allais paradox and the certainty effect for thirty years, but the certainty effect would figure in my decision-making if I were presented with these gambles. Most people are risk-averse, so if we can't take risks or wait for a larger gain in the future because of our anxiety or fear about uncertainty, one obvious solution would be to boost our confidence about the future.

One study concretely illustrates how that can be done. Participants were divided into two groups. One group was told to describe a situation when they lacked power, for instance, a time when their boss made them work over the weekend, or they had to drop out of a tournament they'd trained for because of a sprained ankle. The second group of participants was told to describe a situation when they had power. A participant might write about what it was like when she was captain of a varsity team, with the power to decide training plans and what everyone ate for team dinners. Another might recall working as a manager at a store, with the power to assign tasks for employees. The study found that those who wrote about the situation where they had power were more willing to wait for better rewards than those who imagined situations in which they lacked power.

The pandemic left all of us with anxieties and uncertainties about the future, and we still feel a lack of control.

But even when we are not dealing with an epic disaster, we can feel stuck and helpless at times. To restore our faith in the future, it can be useful to remind ourselves of the times when we had the power to make a difference in our own life or others' lives. This can help us make better choices, ones that are based on facts rather than fears.

Psychological Distance

Another explanation for why we discount the value of things happening in the future may sound obvious: it is because the future simply *feels* distant. As obvious as that might be, it suggests a solution that is anything but.

Let's use spatial distance as an analogy for temporal difference. When there's a fire on your block, even if there's no danger of it spreading to your house, it's shocking. But if there's a fire in another city, you might not even read about it in the news. Here's a more cheerful example. If someone you went to high school with won an Oscar, you would celebrate and feel proud, even though you had nothing to do with it. But if that same Oscar went to someone in a different country, you'd hardly care unless you were a particular fan of that performer. We feel a similar lack of interest about the future, and as a result we discount future rewards or pains.

I was once invited to give a talk in Cambridge, United Kingdom, a small conference scheduled six months in the future. I had a minor surgery scheduled for a month before the conference, but my doctor told me that most people who get this surgery are able to travel within a month. I assumed I would be like most people, and that even if I wasn't, the pain wouldn't be too terrible, so I gladly accepted the invitation. Everything, including the possible pain I might experience five months hence, seemed blurry and hazy. Then, right after the surgery, I realized that I would have to prepare for the talk while I was recovering and still in pain. When I accepted the invitation six months before, I had neglected to consider all the details. I should have known better: I make a point of inviting hard-to-get speakers months in advance when I am planning conferences, because I know that the further off an event is, the more likely they are to commit to it. And here I'd fallen for the same trick myself.

Temporal discounting is why we overcommit ourselves. We grossly underestimate the potential costs, pains, efforts, and time that our commitments will require when they are a long way off. It's not just pains of various kinds that we discount when they are temporally distant; we do the same thing with rewards. Take climate change again. A study revealed that people prefer twenty-one days of improved air quality this year over thirty-five days of improved air quality one year from now. It's easy to imagine how much our present self can enjoy the fresh air, but it is hard to imagine who our future self will be and how much fresh air would be worth to that person.

Is there anything we can do to avoid the pitfalls of psychological distance? One method that has been shown to work is to think about future events with as much specific detail as we can summon in order to make the future feel more real. And there are cool new tools that can help us do this.

In one study, researchers used immersive virtual reality to help young people prepare for their financial futures. First, digitalized avatars of the undergraduates participating in the experiment were created. Then, some of the avatars were altered so they appeared to be close to retirement age. It turned out that the students whose avatars were age-progressed were about twice as likely to allocate a hypothetical $1,000 windfall to their retirement than those who only saw their same-age avatars.

Not many of us have access to sophisticated virtual reality equipment, but simply imagining positive future events can help. In one study, participants were presented with the standard delayed choice situations in which a smaller reward right now—say, 20 euros immediately—was pitted against a larger amount on a later date—say, 35 euros forty-five days later. But before they were asked to make their choices, participants were first instructed to list events they had planned over the next seven months. For instance, Audrey might say she planned to take a holiday in Rome in forty-five days. Then, when they were given a choice between the two rewards, the delayed option was tied to that event. So, Audrey was informed that she could have 20 euros right now, or 35 euros forty-five days later, with the wording HOLIDAY IN ROME written beneath the second option. Reminding people about planned future events significantly reduced irrational discounting of future rewards and encouraged them to choose delayed gratification.

This kind of technique has become pivotal in developing methods to help people reduce their use of tobacco and alcohol and cut their caloric intake. In one study, overweight women participated in afternoon experiment sessions that were timed to occur long after lunch at a time when they would be hungry. First, they were primed with thoughts of foods most people find comforting, such as meatballs, fries, sausages, cookies, and dips, to trigger impulsive eating. Then they were given unlimited access to those foods for fifteen minutes, and asked to rate how good they tasted. During the tasting test, half of the women, chosen at random, listened to audio recordings of their own musings on good things that could happen to them in the future. The other half also listened to an audio recording of their own voice, but it was about a woman writer's travel blog that took place recently, having nothing to do with the participants' own futures. After the fifteen minutes allotted for eating were up, the caloric intake of each participant was measured. Those who were thinking about their future selves took in about 800 calories on average, while those who did not consumed about 1,100.

TO PERSIST OR NOT TO PERSIST

I started this chapter explaining why people's propensity to discount future rewards can be irrational, and discussed several factors that contribute to it so we can try to overcome it. Before closing, I want to add an important caveat. Throughout this discussion, it might have sounded as if resisting immediate rewards and sacrificing for the future is an unequivocally good thing. The idea that everyone can change themselves for the better, and that hard work and grit are more important than innate talent, are staples of popular psychology these days; there are any number of bestselling books about famous people who, despite not showing much talent at the beginning of their careers, achieved great things because they had strength of character and persisted against the odds. Many programs, some of them funded by the government, are aimed at character-building and improving self-control to reduce drug and alcohol abuse and crimes. I applaud these efforts.

At the same time, I'm also afraid that when it is taken to an extreme, a one-sided emphasis on self-control can backfire. Anecdotes about successful people staying the course through thick and thin are always inspiring. But considering only those cases is a perfect example of confirmation bias, as explained in chapter 2. There are also numerous negative examples of people who persisted for years and years in vain. It seems to me that we should moderate our culture of "the little engine that could." Two observations lead me to raise this point.

First, there is an epidemic of anxiety among adolescents and young adults. According to the National Institute of Mental Health, nearly one-third of adolescents have had an anxiety disorder at least once in their lifetime. Anxiety is not only prevalent, it is increasing: anxiety among eighteen- to twenty-five-year-olds increased from 8 percent in 2008 to 15 percent in 2018 (that is, even before the pandemic). I could personally feel that increase. Many brilliant students experience FOMO (fear of missing out)—and not in terms of fun things but crucial steps in their never-ending race for achievement. I was no different. I started out this chapter with the story of how I drove myself to receive a Ph.D. at age twenty-five.

But there's a follow-up to my story. Not long after I received the degree, I took the money I saved from my new postdoctoral fellow job and went to Paris for the first time. Even though I had to stay in a closet-sized room in a youth hostel, everything was shockingly beautiful and delicious. I discovered crêpes and onion soup and learned that it's OK to add as much butter as a thick slice of cheese to a "jambon" (ham) sandwich on a baguette. But the biggest culture shock for me was seeing so many people taking leisurely two-hour lunch breaks with wine on weekdays. I thought lunch was an impediment to being productive, something that you shove into your mouth in ten minutes while staring at a computer or reading an article.

Then, while wandering around museums in Paris and looking at paintings that depicted the strange customs of people who lived two centuries or more ago, I had a thought. The people in the paintings I was looking at thought divorce should be illegal and corsets were indispensable for women's high fashion. What do we take for granted now that future generations will think is not just wrong but ludicrous?

Because I had just recently received my Ph.D., I had been pondering whether all that persistence, sacrifice, and delayed gratification had been worth it. During that trip to the museum, I concluded that perhaps our willingness to live to work might be something that will make people laugh at us in the future. Why did we create a society that not only forces most people to work to live but also makes even the most privileged feel like they have to work desperately hard all the time? We created a mythology that regards the need to reach the top of the mountain because it's there as a measure of human worth, but when we get there, there's always another mountain, and another one after that. Most of us live our whole lives either struggling to stay on solid ground or climbing mountain after mountain.

Actually, it may not take two hundred years for the absurdity of this overemphasis on work to be recognized. Many European countries seem to understand it already. Nordic countries like Denmark, Norway, and Finland rank at the top in the world when it comes to happiness. One reason is free education and health care, which allows for a better work-life balance.

Not only does excessive self-control impede our mental health and happiness, it impairs our physical health, especially among those who are less advantaged in terms of their socioeconomic status. In one study, researchers followed a group of socioeconomically disadvantaged African

American teenagers from rural Georgia for several years. They measured the teenagers' levels of self-control. (Because the results I am about to describe are rather counterintuitive, for skeptical readers, it's worth explaining a bit of details on how this was done: their levels of self-control were assessed by their caregivers and also through self-reports where the teenagers answered how much they agreed with statements like "I usually keep track of my progress toward my goals" and "If I wanted to change, I am confident that I could do it.") As one can easily imagine, these teenagers varied in how much self-control they had. The researchers found that those who showed better self-control at ages seventeen to nineteen showed lower rates of substance abuse and aggressive behavior by age twenty-two. So it was the results we would have predicted, showing the typical benefits of self-control. But the study also reports a surprising discovery: the greater their self-control in mid-adolescence, the *more* they showed signs of immune cell aging by young adulthood. Another study also found the similarly startling pattern: low-socioeconomic-status children with better self-control showed *greater* cardiometabolic risks (as indicated by obesity, blood pressure, and stress hormones), despite having fewer law-breaking behaviors and less substance use.

What is going on? When these disadvantaged but self-disciplined teenagers start doing well at school and in life, they want to maintain that level or do even better. But because they are in disadvantaged environments, they are continually barraged with challenges and difficulties. Because of their high level of self-regulation, they fight against those challenges rather than giving up. It's like being in a never-ending battle for years. Their stress hormone systems are continuously activated, to the eventual detriment of their physical health.

The detrimental effects of excessive self-control do not appear to be limited to disadvantaged children. In another study, undergraduates who were not necessarily disadvantaged were recruited to participate in psychology experiments in return for partial course credits. Their preexisting desire to have self-control was measured by asking them how much they agreed with statements such as "I want to have my control over my feelings" and "I wish I had a better ability to change unwanted habits."

Then, all of the participants were asked to work on a copying task. For some of them, the task was really simple; it was just a matter of using a keyboard to copy a paragraph in their native language, which was Hebrew. For the others, the task was heinous: they had to copy a paragraph in a foreign language, which was English, while typing only with their nondominant hand, skipping the letter "e," and not using the spacebar, so that "If a cluttered desk is a sign of a cluttered mind, of what, then, is an empty desk a sign?" (Albert Einstein) would be copied as "Ifacluttrddskisasigno-facluttrdmind,ofwhat,thn,isanmptydskasign?" (Wow, that was hard to type even with both hands and in the language I use every day.)

Now, we would think that those who fervently valued

self-control would do better at both of these tasks, right? Nope. A strong desire for self-control showed some benefit for simpler tasks, but the opposite happened with difficult tasks: those with a high desire for self-control performed worse than those with a low desire.

Why did that happen? Because the difficult task required an extreme level of self-control. The people with a strong desire for self-control would have quickly realized the gap between their aspiration (that is, being perfect!) and their actual performance. When their goal appears to be unreachable, they feel discouraged. As a result, they put in less effort and end up performing worse than they could have.

I wonder whether something like this may at least partly explain the elevated levels of anxiety among young people. Those in disadvantaged environments feel they should do much better than how they started out. Those in privileged environments are surrounded by stellar students and constantly exposed to social media postings in which others advertise their best talents and achievements, incessantly reminding them of the aspirational level that they "ought to" reach. The disparity they feel between their actual selves and their ideal selves can make these highly self-regulated students push themselves too hard, creating stress, anxiety, and a feeling of defeat.

It is no small task to know when to persist and when to quit. To that end, I remind myself every day to enjoy the process of doing without jumping ahead to results. I listen to my yoga instructor when she says "Breathe" when I'm trying to do the camel pose, kneeling with my thighs perpendicular to the floor, bending my spine backwards so that my chest is facing the ceiling while my hands are trying to grab my heels that are still far, far away. As my instructor says, breathing should be your guide for how hard you can push yourself: if you cannot breathe easily, don't do it. I swear by that advice, which must have staved off countless occasions for harm that I, a control freak, could have easily inflicted on myself. I may never be able to achieve the camel pose, but I can blame that on my short arms. I still enjoy the feeling of my spine waking up and the blood rushing through my head while I maintain my breathing.

If a goal is worth pursuing, even the pain that accompanies our practice feels good—just like the pain of good exercise, spicy hot pot, or icy cold tingling soda. But if you feel like you're hurting yourself to achieve rewards and the only thing you enjoy is the final goal and not the process, it's probably time to rethink—not just about your priorities but the way you think about them.

EPILOGUE

WHY DO PEOPLE WANT TO BECOME better at thinking? One candid answer I've heard several times goes something like this: "Because I want to learn how to outsmart everyone in the room." Understanding loss aversion, for instance, may help you devise business or investment strategies that capitalize on others' fear. Learning that people come up with widely different interpretations to explain the same outcome according to the order in which information is presented to them can be useful if your aim is to manipulate others' opinions. I hope you won't use the book in this way.

I've long wondered how cognitive psychology can make the world a better place. Outsmarting or defeating over others isn't the best way to create a better world. So, let's loop back and see how understanding errors in thinking can make the world a better place. I believe a better world is a fairer one, and in order to be fair, we need more unbiased thinking.

For starters, each of us should be fair to ourselves. We shouldn't be underconfident which can happen when we selectively search for reasons to perpetuate our insecurity (chapter 2) or use all of our creative energy to come up with the worst possible interpretations of our misfortunes (chapter 6). It's also not fair to ourselves to be overconfident, ignoring our limitations and putting ourselves in situations that we can't handle (chapter 1). The decisions we make for ourselves should be as impartial as possible and based on statistical principles and probability theories, because they provide the most accurate predictions (chapter 4). Knowing how we can fall prey to anecdotes, framing, and loss aversion allows us to outsmart people who try to outsmart *us* by exploiting those techniques (chapter 5). We're not being fair to ourselves if we don't sufficiently take our future into consideration, but it is equally unfair to sacrifice our present for the future (chapter 8).

And we should be fairer to others—and better thinking is fairer because it's less biased. If you want to claim that a group of people are special because they are good at something, it's never enough to show that they are good at that thing, because a different group of people could be also good or even better at it. Providing equal opportunities to everyone is the only proper way of testing such a hypothesis (chapter 2). Once we realize there are always multiple possible causes for an event, we can assess credit as well as blame more fairly (chapter 3). And the road to a more equitable society is much more direct when we ask people what they need and want instead of assuming that we know already (chapter 7). When we can anticipate others' shortcomings, such as the ubiquitous planning fallacy (chapter 1), and have a Plan B in place, we can be more patient with them—especially those who haven't read this book!

Just like a new pair of jeans or shoes, it does take time to break in new ways of thinking. We certainly won't and can't fix everything, but it also won't hurt to spend a bit more time, individually and together, talking about how we're doing and sharing our own thoughts with one another.

ACKNOWLEDGMENTS

First, I would like to thank all the cognitive psychologists whose work created the foundation for this book, especially those whose studies I cited. In particular, I believe the world would have been much worse off without Daniel Kahneman and the late Amos Tversky, and I cannot thank them enough for their groundbreaking research.

I am also enormously indebted to all the students who took my "Thinking" course. Their eagerness to learn and their willingness to laugh at their errors have inspired me to spend more than twenty hours a week preparing for three-hour-long lectures, searching for better and more up-to-date examples and jokes to keep them engaged and to help the lessons linger a bit longer in their minds. This book would not have been possible without their enthusiasm. I would like to especially thank Alicia Mazzurra, who took the course in the fall of 2021, for the subtitle of this book.

Will Schwalbe at Flatiron, a genius storyteller and veteran editor, patiently and skillfully walked me through multiple versions of the manuscript. He has the highest level of "theory of mind" that I have seen, clearly understanding not only the writer's challenges but the readers' perspectives. I have enjoyed working with this brilliant editor so much that I am almost sorry that the book is nearly finished.

Jim Levine, my literary agent, helped me especially during the initial phase, when the book was being conceptualized. I am grateful that he insisted that I stick with the positive thesis of how to improve thinking rather than the negative thesis of what is wrong with our thinking. Arthur Goldwag significantly improved my prose throughout, editing my non-native speaker's English while maintaining my voice. Samantha Zukergood and Andrea Mosqueda at the editor's office offered the younger generation's perspective. I also thank Bill Warhop, the copy editor at Flatiron, for his thorough work.

My own research presented throughout the book was supported by grants from the National Institute of Mental Health and the National Human Genome Research Institute, and a generous gift from the Reboot Foundation.

Finally, my husband, Marvin Chun. When I was an assistant professor at Yale around 1998, I attended a panel session for female professors on how to have everything, namely career and family. One panelist said there is only one secret: find the right husband. Fortunately, I already had. Throughout our marriage, we split the housework, child-rearing, and our children's last names. He has always supported my career, becoming genuinely distressed whenever I lose confidence in myself. As a cognitive psychologist who taught highly popular "Introduction to Psychology" courses for years, he went through the first draft of every chapter, providing constructive and critical suggestions. As a husband, he also had to live with all my bragging and whining as I experienced the ups and downs of writing while we were confined to our house during the pandemic. Thank you for everything.

NOTES

1: The Allure of Fluency

9 **a study on the illusion of fluency that can occur when we are learning new skills:** Michael Kardas and Ed O'Brien, "Easier seen than done: Merely watching others perform can foster an illusion of skill acquisition," *Psychological Science* (2018).

12 **people are more willing to derive a cause from a correlation:** Woo-kyoung Ahn and Charles W. Kalish, "The role of mechanism beliefs in causal reasoning," *Explanation and Cognition* (2000): 199–225.

15 **people's expectations for their performance in the market:** Adam L. Alter and Daniel M. Oppenheimer, "Predicting short-term stock fluctuations by using processing fluency," *Proceedings of the National Academy of Sciences* 103, no. 24 (2006): 9369–72.

16 **it makes people think they are more knowledgeable than they really are:** Matthew Fisher, Mariel K. Goddu, and Frank C. Keil, "Searching for explanations: How the internet inflates estimates of internal knowledge," *Journal of Experimental Psychology: General* 144, no. 3 (2015): 674.

24 **spelling out one's knowledge can reduce overconfidence:** Leonid Rozenblit and Frank Keil, "The misunderstood limits of folk science: An illusion of explanatory depth," *Cognitive Science* 26, no. 5 (2002): 521–62.

25 **One study showed that it can reduce political extremism:** Philip M. Fernbach, Todd Rogers, Craig R. Fox, and Steven A. Sloman, "Political extremism is supported by an illusion of understanding," *Psychological Science* 24, no. 6 (2013): 939–46.

29 **One study that examined the planning fallacy:** Roger Buehler and Dale Griffin, "Planning, personality, and prediction: The role of future focus in optimistic time predictions," *Organizational Behavior and Human Decision Processes* 92, no. 1–2 (2003): 80–90.

33 **studies done with nonhuman animals like birds and rats:** Stephanie M. Matheson, Lucy Asher, and Melissa Bateson, "Larger, enriched cages are associated with 'optimistic' response biases in captive European starlings (Sturnus vulgaris)," *Applied Animal Behaviour Science* 109, no. 2–4 (2008): 374–83.

2: Confirmation Bias

43 **In Wason's first experiment with the 2–4–6 task:** Peter C. Wason, "On the failure to eliminate hypotheses in a conceptual task," *Quarterly Journal of Experimental Psychology* 12, no. 3 (1960): 129–40.

48 **One of the questions I use, taken from:** Keith E. Stanovich, Richard F. West, and Maggie E. Toplak, *The rationality quotient: Toward a test of rational thinking* (CITY: MIT Press, 2016).

52 **more than twenty-six million people:** A. Regalado, "More than 26 million people have taken an at-home ancestry test," *MIT Technology Review,* February 11, 2019, www.technologyreview .com/2019/02/11/103446/more-than-26-million-peoplehave -taken-an-at-home-ancestry-test/.

53 **to make sense of themselves in light of their genetic test results:** Matthew S. Lebowitz and Woo-kyoung Ahn, "Testing positive for a genetic predisposition to depression magnifies retrospective memory for depressive symptoms," *Journal of Consulting and Clinical Psychology* 85, no. 11 (2017): 1052.

67 **notorious 2–4–6 problem when it was framed as discovering two rules:** Ryan D. Tweney, Michael E. Doherty, Winifred J. Worner, Daniel B. Pliske, Clifford R. Mynatt, Kimberly A. Gross, and Daniel L. Arkkelin, "Strategies of rule discovery in an inference task," *Quarterly Journal of Experimental Psychology* 32, no. 1 (1980): 109–23.

69 **participants in a study ended up rating themselves to be significantly unhappier:** Ziva Kunda, Geoffrey T. Fong, Rasyid Sanitioso, and Emily Reber, "Directional questions direct self-conceptions," *Journal of Experimental Social Psychology* 29, no. 1 (1993): 63–86.

71 **The researchers behind the Mozart effect reported:** Frances H. Rauscher, Gordon L. Shaw, and Katherine N. Ky. "Music and spatial task performance." *Nature* 365, no. 6447 (1993): 611–611.

71 **One study examined whether one of these bestselling videos:** Judyf S. DeLoache, Cynthia Chiong, Kathleen Sherman, Nadia

Islam, Mieke Vanderborght, Georgene L. Troseth, Gabrielle A. Strouse, and Katherine O'Doherty. "Do babies learn from baby media?" *Psychological Science* 21, no. 11 (2010): 1570–1574.

3: The Challenge of Causal Attribution

76 **if only Wilson had not caught the flu, there would have been no Holocaust:** For detailed accounts, see for example John M. Barry, *The great influenza: The story of the deadliest pandemic in history* (New York: Viking Press, 2004).

84 **a study showed that when people received a short-term monetary bonus:** Liad Bareket-Bojmel, Guy Hochman, and Dan Ariely, "It's (not) all about the Jacksons: Testing different types of short-term bonuses in the field," *Journal of Management* 43, no. 2 (2017): 534–54.

86 **Such unwarranted discounting has devastating real-life consequences:** Ilan Dar-Nimrod and Steven J. Heine, "Exposure to scientific theories affects women's math performance," *Science* 314, no. 5798 (2006): 435.

91 **to blame actions more than inactions:** Daniel Kahneman and Amos Tversky, "The psychology of preferences," *Scientific American* 246, no. 1 (1982): 160–73.

94 **to most people, temporal order matters:** Dale T. Miller and Saku Gunasegaram, "Temporal order and the perceived mutability of events: Implications for blame assignment," *Journal of Personality and Social Psychology* 59, no. 6 (1990): 1111.

96 **Our propensity to assign blame:** Vittorio Girotto, Paolo Legrenzi, and Antonio Rizzo, "Event controllability in counterfactual thinking," *Acta Psychologica* 78, no. 1–3 (1991): 111–33.

98 **rumination can cause depression:** Sonja Lyubomirsky and Susan Nolen-Hoeksema, "Effects of self-focused rumination on negative thinking and interpersonal problem solving," *Journal of Personality and Social Psychology* 69, no. 1 (1995): 176–90.

100 **rumination actually prevents us from effectively solving our problems:** Susan Nolen-Hoeksema, Susan, Blair E. Wisco, and Sonja Lyubomirsky, "Rethinking rumination," *Perspectives on Psychological Science* 3, no. 5 (2008): 400–24.

100 **one that demonstrated the effectiveness of a self-distanced approach:** Ethan Kross, Ozlem Ayduk, and Walter Mischel,

"When asking 'why' does not hurt distinguishing rumination from reflective processing of negative emotions," *Psychological Science* 16, no. 9 (2005): 709–15.

101 **Self-distancing also had a long-term benefit:** Ethan Kross and Ozlem Ayduk, "Facilitating adaptive emotional analysis: Distinguishing distanced-analysis of depressive experiences from immersed-analysis and distraction," *Personality and Social Psychology Bulletin* 34, no. 7 (2008): 924–38.

4: The Perils of Examples

107 **The campaign increased quit attempts by 12 percent:** Tim McAfee, Kevin C. Davis, Robert L. Alexander Jr, Terry F. Pechacek, and Rebecca Bunnell, "Effect of the first federally funded US antismoking national media campaign," *The Lancet* 382, no. 9909 (2013): 2003–11.

111 **One study used undergraduate students to gain insights:** Eugene Borgida and Richard E. Nisbett, "The differential impact of abstract vs. concrete information on decisions," *Journal of Applied Social Psychology* 7, no. 3 (1977): 258–71.

112 **if they became aware of this reasoning fallacy:** Deborah A. Small, George Loewenstein, and Paul Slovic, "Sympathy and callousness: The impact of deliberative thought on donations to identifiable and statistical victims," *Organizational Behavior and Human Decision Processes* 102, no. 2 (2007): 143–53.

114 **there *is* a way to help people seek out more data:** Geoffrey T. Fong, David H. Krantz, and Richard E. Nisbett, "The effects of statistical training on thinking about everyday problems," *Cognitive Psychology* 18, no. 3 (1986): 253–92.

127 **In a study conducted in the early 1980s:** David M. Eddy, "Probabilistic reasoning in clinical medicine: Problems and opportunities," *Judgment under Uncertainty: Heuristics and Biases*, edited by Daniel Kahneman, Paul Slovic, and Amos Tversky (Cambridge: Cambridge University Press, 1982), 249–67.

128 **here is how he could have done it using his own equation:** Philip Dawid and Donald Gillies, "A Bayesian analysis of Hume's argument concerning miracles," *Philosophical Quarterly (1950-)* 39, no. 154 (1989): 57–65.

129 **According to a 2017 report by the U.S. Government Accountability Office:** United States Government Accountability Office Report to Congressional Requesters, "Countering violent extremism: Actions needed to define strategy and assess progress of federal efforts," (GAO-17-300I), April 2017, https://www.gao.gov/products/gao-17-300; I would like to thank my former undergraduate student Alexandra Otterstrom for pointing me to the sources that this analysis is based on.

133 **In a study that used these two problems:** Mary L. Gick and Keith J. Holyoak, "Schema induction and analogical transfer," *Cognitive Psychology* 15, no. 1 (1983): 1–38.

5: Negativity Bias

138 **the ways that positive and negative reviews affect sales:** Geng Cui, Hon-Kwong Lui, and Xiaoning Guo, "The effect of online consumer reviews on new product sales," *International Journal of Electronic Commerce* 17, no. 1 (2012): 39–58.

139 **people give more weight to negative behavior:** Susan T. Fiske, "Attention and weight in person perception: The impact of negative and extreme behavior," *Journal of Personality and Social Psychology* 38, no. 6 (1980): 889–906.

139 **Negative events also affect our lives more than positive events:** Roy F. Baumeister, Ellen Bratslavsky, Catrin Finkenauer, and Kathleen D. Vohs, "Bad is stronger than good," *Review of General Psychology* 5, no. 4 (2001): 323–70.

140 **researchers cooked ground beef:** Irwin P. Levin and Gary J. Gaeth, "How consumers are affected by the framing of attribute information before and after consuming the product," *Journal of Consumer Research* 15, no. 3 (1988): 374–78.

141 **negativity bias in the context of college admission processes:** Woo-kyoung Ahn, Sunnie S. Y. Kim, Kristen Kim, and Peter K. McNally, "Which grades are better, A's and C's, or all B's? Effects of variability in grades on mock college admissions decisions," *Judgment & Decision Making* 16, no. 6 (2019): 696–710.

146 **one of the most important papers in behavioral economics:** Daniel Kahneman and Amos Tversky, "Prospect theory: An analysis of decision under risk," *Econometrica* 47, no. 2 (1979): 263–92.

149 **participants were told to imagine one of the two situations:** C. Whan Park, Sung Youl Jun, and Deborah J. MacInnis, "Choosing what I want versus rejecting what I do not want: An application of decision framing to product option choice decisions," *Journal of Marketing Research* 37, no. 2 (2000): 187–202.

150 **carried out what they called "field experiments":** Roland G. Fryer, Steven D. Levitt, John List, and Sally Sadoff, *Enhancing the efficacy of teacher incentives through loss aversion: A field experiment,* No. w18237, National Bureau of Economic Research, 2012.

153 **undergraduates were given a choice between a mug:** Jack L. Knetsch, "The endowment effect and evidence of nonreversible indifference curves," *American Economic Review* 79, no. 5 (1989): 1277–84.

154 **the pain of losses is literally physical:** C. Nathan DeWall, David S. Chester, and Dylan S. White, "Can acetaminophen reduce the pain of decision-making?," *Journal of Experimental Social Psychology* 56 (2015): 117–20.

158 **it can literally be a matter of life and death:** Barbara J. McNeil, Stephen G. Pauker, Harold C. Sox Jr., and Amos Tversky, "On the elicitation of preferences for alternative therapies," *New England Journal of Medicine* 306, no. 21 (1982): 1259–62.

158 **we can also try to reframe the questions we ask ourselves:** Eldar Shafir, "Choosing versus rejecting: Why some options are both better and worse than others," *Memory & Cognition* 21, no. 4 (1993): 546–56.

6: Biased Interpretation

163 **Babies who slept with a light on:** Graham E. Quinn, Chai H. Shin, Maureen G. Maguire, and Richard A. Stone, "Myopia and ambient lighting at night," *Nature* 399, no. 6732 (1999): 113–14.

163 **As CNN summarized it:** "Night-light may lead to nearsightedness," CNN.com, May 13, 1999, http://www.cnn.com/HEALTH/9905/12/children.lights/.

164 **debunking this earlier study:** Karla Zadnik, Lisa A. Jones, Brett C. Irvin, Robert N. Kleinstein, Ruth E. Manny, Julie A. Shin, and Donald O. Mutti, "Myopia and ambient night-time lighting," *Nature* 404, no. 6774 (2000): 143–44.

164 **CNN duly corrected its earlier report:** Ulysses Torassa, "Leave it on: Study says night lighting won't harm children's eyesight," CNN.com, March 8, 2000, https://www.cnn.com/2000 /HEALTH/children/03/08/light.myopia.wmd/index.html.

166 **that belief is imprinted and does not get revised:** Eric G. Taylor and Woo-kyoung Ahn, "Causal imprinting in causal structure learning," *Cognitive Psychology* 65, no. 3 (2012): 381–413.

169 **what happens when two candidates for a research job are identical:** Corinne A. Moss-Racusin, John F. Dovidio, Victoria L. Brescoll, Mark J. Graham, and Jo Handelsman, "Science faculty's subtle gender biases favor male students," *Proceedings of the National Academy of Sciences* 109, no. 41 (2012): 16474–79.

170 **Participants . . . were asked to play a video game:** Joshua Correll, Bernadette Park, Charles M. Judd, and Bernd Wittenbrink, "The police officer's dilemma: Using ethnicity to disambiguate potentially threatening individuals," *Journal of Personality and Social Psychology* 83, no. 6 (2002): 1314–29.

172 **A seminal study, published in 1979:** Charles G. Lord, Lee Ross, and Mark R. Lepper, "Biased assimilation and attitude polarization: The effects of prior theories on subsequently considered evidence," *Journal of Personality and Social Psychology* 37, no. 11 (1979): 2098–109.

176 **individuals at different levels of quantitative reasoning skills:** Dan M. Kahan, Ellen Peters, Erica Cantrell Dawson, and Paul Slovic, "Motivated numeracy and enlightened self-government," *Behavioural Public Policy* 1, no. 1 (2017): 54–86.

183 **can people see an identical object in two opposite ways:** Jessecae K. Marsh and Woo-kyoung Ahn, "Spontaneous assimilation of continuous values and temporal information in causal induction," *Journal of Experimental Psychology: Learning, Memory, and Cognition* 35, no. 2 (2009): 334–52.

7: The Dangers of Perspective-Taking

196 **participants were tested on their sarcasm:** Justin Kruger, Nicholas Epley, Jason Parker, and Zhi-Wen Ng, "Egocentrism over e-mail: Can we communicate as well as we think?," *Journal of Personality and Social Psychology* 89, no. 6 (2005): 925–36.

197 **This study used ambiguous sentences:** Kenneth Savitsky, Boaz Keysar, Nicholas Epley, Travis Carter, and Ashley Swanson, "The closeness-communication bias: Increased egocentrism among friends versus strangers," *Journal of Experimental Social Psychology* 47, no. 1 (2011): 269–73.

199 **even college students have similar difficulties:** Susan A. J. Birch and Paul Bloom, "The curse of knowledge in reasoning about false beliefs," *Psychological Science* 18, no. 5 (2007): 382–86.

201 **120 songs were tapped and only three were correctly guessed:** L. Newton, "Overconfidence in the communication of intent: Heard and unheard melodies" (unpublished Ph.D. diss., Stanford University, 1990).

203 **An example of this phenomenon:** Stephen M. Garcia, Kimberlee Weaver, and Patricia Chen, "The status signals paradox," *Social Psychological and Personality Science* 10, no. 5 (2019): 690–96.

205 **people forget to take other people's perspectives into account:** Shali Wu and Boaz Keysar, "The effect of culture on perspective taking," *Psychological Science* 18, no. 7 (2007): 600–606.

210 **children in this age cohort can be taught to understand false beliefs:** Xiao Pan Ding, Henry M. Wellman, Yu Wang, Genyue Fu, and Kang Lee, "Theory-of-mind training causes honest young children to lie," *Psychological Science* 26, no. 11 (2015): 1812–21.

213 **Let me illustrate this more concretely:** Claire L. Adida, Adeline Lo, and Melina R. Platas, "Perspective taking can promote short-term inclusionary behavior toward Syrian refugees," *Proceedings of the National Academy of Sciences* 115, no. 38 (2018): 9521–26.

214 **A team of three researchers:** Tal Eyal, Mary Steffel, and Nicholas Epley, "Perspective mistaking: Accurately understanding the mind of another requires getting perspective, not taking perspective," *Journal of Personality and Social Psychology* 114, no. 4 (2018): 547–71.

219 **tested whether we can better discern:** David Comer Kidd and Emanuele Castano, "Reading literary fiction improves theory of mind," *Science* 342, no. 6156 (2013): 377–80.

219 **evaluated the replicability of social science experiments:** Colin F. Camerer, Anna Dreber, Felix Holzmeister, Teck-Hua

Ho, Jürgen Huber, Magnus Johannesson, Michael Kirchler et al., "Evaluating the replicability of social science experiments in *Nature* and *Science* between 2010 and 2015," *Nature Human Behaviour* 2, no. 9 (2018): 637–44.

8: The Trouble with Delayed Gratification

225 **The answer is no:** The discussion of irrationality of delay discounting is based on Jonathan Baron, *Thinking and deciding* (Cambridge: Cambridge University Press, 2000).

230 **One of the earliest studies on delayed gratification:** Walter Mischel, Ebbe B. Ebbesen, and Antonette Raskoff Zeiss, "Cognitive and attentional mechanisms in delay of gratification," *Journal of Personality and Social Psychology* 21, no. 2 (1972): 204–18.

231 **a subsequent study debunked the marshmallow test:** Tyler W. Watts, Greg J. Duncan, and Haonan Quan, "Revisiting the marshmallow test: A conceptual replication investigating links between early delay of gratification and later outcomes," *Psychological Science* 29, no. 7 (2018): 1159–77.

231 **that follow-up study was later convincingly critiqued:** See for example Armin Falk, Fabian Kosse, and Pia Pinger, "Re-revisiting the marshmallow test: a direct comparison of studies by Shoda, Mischel, and Peake (1990) and Watts, Duncan, and Quan (2018)," *Psychological Science* 31, no. 1 (2020): 100–104.

232 **Distractions can help pigeons delay gratification:** James Grosch and Allen Neuringer, "Self-control in pigeons under the Mischel paradigm," *Journal of the Experimental Analysis of Behavior* 35, no. 1 (1981): 3–21.

233 **the feeling of uncertainty can mess up our judgments:** Amos Tversky and Eldar Shafir, "The disjunction effect in choice under uncertainty," *Psychological Science* 3, no. 5 (1992): 305–10.

240 **one obvious solution would be to boost our confidence about the future:** Priyanka D. Joshi and Nathanael J. Fast, "Power and reduced temporal discounting," *Psychological Science* 24, no. 4 (2013): 432–38.

242 **people prefer twenty-one days of improved air quality:** David J. Hardisty and Elke U. Weber, "Discounting future green: money versus the environment," *Journal of Experimental Psychology: General* 138, no. 3 (2009): 329–40.

243 **researchers used immersive virtual reality:** Hal E. Hershfield, Daniel G. Goldstein, William F. Sharpe, Jesse Fox, Leo Yeykelis, Laura L. Carstensen, and Jeremy N. Bailenson, "Increasing saving behavior through age-progressed renderings of the future self," *Journal of Marketing Research* 48, no. SPL (2011): S23–37.

243 **simply imagining positive future events can help:** Jan Peters and Christian Büchel, "Episodic future thinking reduces reward delay discounting through an enhancement of prefrontal-mediotemporal interactions," *Neuron* 66, no. 1 (2010): 138–48.

244 **overweight women participated in afternoon experiment sessions:** T. O. Daniel, C. M. Stanton, and L. H. Epstein, "The future is now: Reducing impulsivity and energy intake using episodic future thinking," *Psychological Science* 24, no. 11 (2013): 2339–42.

246 **Anxiety is not only prevalent, it is increasing:** Renee D. Goodwin, Andrea H. Weinberger, June H. Kim, Melody Wu, and Sandro Galea, "Trends in anxiety among adults in the United States, 2008–2018: Rapid increases among young adults," *Journal of Psychiatric Research* 130 (2020): 441–46.

247 **it impairs our physical health:** Gregory E. Miller, Tianyi Yu, Edith Chen, and Gene H. Brody, "Self-control forecasts better psychosocial outcomes but faster epigenetic aging in low-SES youth," *Proceedings of the National Academy of Sciences* 112, no. 33 (2015): 10325–30.

247 **researchers followed a group of socioeconomically disadvantaged:** Gene H. Brody, Tianyi Yu, Edith Chen, Gregory E. Miller, Steven M. Kogan, and Steven R. H. Beach, "Is resilience only skin deep? Rural African Americans' socioeconomic status–related risk and competence in preadolescence and psychological adjustment and allostatic load at age 19," *Psychological Science* 24, no. 7 (2013): 1285–93.

249 **undergraduates who were not necessarily disadvantaged were recruited to participate:** Liad Uziel and Roy F. Baumeister, "The self-control irony: Desire for self-control limits exertion of self-control in demanding settings," *Personality and Social Psychology Bulletin* 43, no. 5 (2017): 693–705.

INDEX

ABOUT THE AUTHOR

WOO-KYOUNG AHN is the John Hay Whitney Professor of Psychology at Yale University. After receiving her Ph.D. in psychology from the University of Illinois Urbana-Champaign, she was assistant professor at Yale University and associate professor at Vanderbilt University. In 2022, she received Yale's Lex Hixon Prize for teaching excellence in the social sciences. Her research on thinking biases has been funded by the National Institutes of Health, and she is a fellow of the American Psychological Association and the Association for Psychological Science.